MW01349304

The Holstein Cattle Herd Book

A Record of Holstein Cattle in America

by The Assoc. of Breeders of Thoroughbred Holstein Cattle

with an introduction by Jackson Chambers

Self Reliance Books

Get more historic titles on animal and stock breeding, gardening and old fashioned skills by visiting us at:

http://selfreliancebooks.blogspot.com/

Introduction

I am pleased to present another title in the "Cattle" series.

The work is in the Public Domain and is re-printed here in accordance with Federal Laws.

As with all reprinted books of this age that are intended to perfectly reproduce the original edition, considerable pains and effort had to be undertaken to correct fading and sometimes outright damage to existing proofs of this title. At times, this task is quite monumental, requiring an almost total "rebuilding" of some pages from digital proofs of multiple copies. Despite this, imperfections still sometimes exist in the final proof and may detract from the visual appearance of the text.

I hope you enjoy reading this book as much as I enjoyed making it available to readers again.

Jackson Chambers

Association of Breeders of Thoroughbred Holstein Cattle.

OFFICERS FOR 1872-73.

President.

WINTHROP W. CHENERY.

Vice-Presidents.

WILLIAM A. RUSSELL. C. C. WALWORTH. THOMAS B. WALES, Jr.

Secretary and Treasurer.

CHARLES HOUGHTON.

ANNUAL MEETING, FIRST WEDNESDAY IN MARCH.

At a meeting of THE ASSOCIATION OF BREEDERS OF THOROUGHBRED HOLSTEIN CATTLE, held March 15th, 1871, the following, amongst other resolutions, were passed unanimously:—

Resolved, That it is expedient and desirable that a *Holstein Herd Book* be published by authority of this Association, containing the pedigrees of all animals approved by the Committee on Pedigrees.

Resolved, That this Association will deem no animal to be Thoroughbred Holstein, except those large, improved black-and-white cattle imported from the provinces of North Holland, Holstein, or intermediate territory; or which cannot be traced in direct line, on the side of both sire and dam, to animals of undoubted purity of blood of said importations. And, whereas, these cattle have in this country been variously designated as "Holstein or Dutch," "Dutch," and "Dutch or Holstein," it is hereby

Resolved, That this Association will discountenance all confusion of terms, and recognize *Holstein* as the proper name of this race of cattle.

Resolved, That the President of this Association be authorized, and hereby requested, to prepare for publication a *Holstein Herd Book*, embracing a sketch of the Holstein race of cattle.

Attest:

CHARLES HOUGHTON,
Secretary.

PREFACE

The undersigned, having been requested by *The Association of Breeders of Thorough-bred Holstein Cattle* to "prepare for publication a Holstein Herd Book, embracing a sketch of the Holstein race of Cattle," presents the following as the result of his labors.

It is believed the Herd Book includes every animal entitled to record. We have endeavored to register every imported animal, and the pedigree of every animal bred in this country, traceable directly to imported stock on the side of both sire and dam, in accordance with the By-Laws and Resolves of the Association.

By a somewhat careful and thorough investigation of the available material upon the subject, we are able to present in the "sketch of the Holstein race of Cattle," a compilation of facts and opinions, which we trust may prove acceptable to the members of the Association.

WINTHROP W. CHENERY.

MAY 20th, 1872.

EARL OF MIDDLESEX. At 3 Years and 10 Months.

PROPERTY OF THE DOYLESTOWN AGRICULTURAL AND MECHANICS INSTITUTE, DOYLESTOWN, BUCKS COUNTY PA.

Bred by Winthrop W. Chenery Belmont, Mass. Color White and Black, Calved March 12th, 1868. Got by Van Tromp, Dam, Zuider Zee. both imported from North Holland. by W. W. Chenery.

HOLSTEIN CATTLE.

The remote origin of the Holstein race of cattle affords a theme for unlimited speculation and discussion. According to tradition, however, as stated by the best authorities, all that is certainly known upon the subject is, that for an indefinite period anterior to the records of history, there existed in the Duchy of Holstein a superior race of cattle, and that thence the finest cattle of the north of Europe have been derived. The present large, improved black-and-white cattle of North Holland, Friesland and Oldenburg, which all possess the same general characteristics, yet present in the different localities some slight dissimilarity, and have perhaps been brought to the highest degree of perfection in the first-named Province, undoubtedly descended from the original stock of Holstein.

In the seventeenth century, as related by the historian Motley, in his History of the United Netherlands, the cattle interest in Holland had become of prime importance to the people, and was in the most thrifty condition. He says: "On that scrap of solid ground, rescued by human energy from the ocean, were the most fertile pastures in the world. On these pastures grazed the most famous cattle in the world. An ox often weighed more than two thousand pounds. The cows produced two and three calves at a time, and the sheep four and five lambs. In a single village, four thousand kine were counted. Butter and cheese were exported to the annual value of a million; salted provisions to an incredible extent. The farmers were industrious, thriving and independent. It is an amusing illustration of the agricultural thrift and republican simplicity of this people that on one occasion a farmer proposed to Prince Maurice that he should marry his daughter, promising with her a dowry of a hundred thousand florins."

An approximate idea of the antiquity of this race of cattle may be deduced from the remark of a French historian, who, writing in 1350 says "that at a certain siege, the besieged could only receive their supply of butter from Holland, which had been famous for its dairy products for five hundred years"; and, as it is known that Holland was indebted to Holstein for its superior breed of dairy cows, it is obvious that the origin of the Holstein cattle must be assigned to a period still more remote.

It is exceedingly difficult to obtain trustworthy information relative to any of the various races and breeds of cattle upon the European Continent, and both English and American authors have in general been very reticent in regard to the improved Holstein cattle of modern times. Seemingly content to devote their attention almost exclusively to those races or breeds of cattle indigenous to or which have been naturalized in the British Islands, English writers have published elaborate histories of the Shorthorn, Jersey, Hereford, Devon, Ayrshire and other breeds of less importance, entirely ignoring the existence of, or dismissing with slight notice, the Holstein, Swiss, Hungarian, and other valuable dairy breeds of Continental Europe; and American writers, it must be said, have, with recent honorable exceptions, copied blindly after the English. The attention of American breeders has thus been directed to the British breeds almost exclusively, and, in their efforts at improvement they have generally been led to look to the British Islands alone for breeds of cattle for importation to this country, without regard to their adaptation to our soil, climate, vegetation, or to the character of our seasons and systems of breeding. Hence the Holstein cattle have not heretofore, or until quite recently, received that appreciation in this country to which they are fairly entitled by reason of their unequalled combination of desirable properties, especially their preëminent dairy qualities, and, also, by their adaptation to the climate and soil of a large portion of the United States.

Dairy husbandry, it is however known, had become a leading feature in the agriculture of Holland and the adjacent provinces at a very early date, and that consequently special attention had been given to the selection and breeding of dairy stock there, long before the English breeders had commenced their efforts at improvement; and, moreover, we learn incidentally not only of the existence of a superior race of dairy cattle in those provinces at an early period, but, also, that the importation of those cattle into the British Islands produced a marked effect and caused great improvement in some of the British breeds.

This view is supported by Wilson, who, in describing the Shorthorn breed, says: "Their origin in Britain belongs to the counties of York and Durham but is very obscurely known. Toward the close of the 17th century, or perhaps at an earlier period, a bull and some cows, which appear to have been one source of the breed, were introduced to Holderness from some part of Continental Europe between Denmark and France. . . . They were better milkers, larger in size, and more capable of being fattened to an enormous bulk than almost any other cattle which were then known; and, on these accounts, they were esteemed, propagated, and intermixed with such of the native cattle as most nearly resembled them;" and again, he says, that "during the latter part of the last century, numerous bulls, which proved another *source of the present Shorthorns,* but in some degree identical with the first, were imported to the counties of York and Durham from *Holstein and Holland*. The frame of the cattle of the present day in *Holstein and Holland*, is superior to that of the old Teeswater breed, and somewhat similar to that of the modern improved Shorthorns. . . . . Improvements of successive stages, but of unrecorded pedigree, were made by crossings of the Teeswater with the *Dutch and the Holstein*, till a new and established breed was produced, called the *Teeswater Shorthorns*."

And that eminent English author, Professor Low, writing in 1840 in relation to the Shorthorn breed, says that at a period "near our own times, it appears that cattle were frequently brought from the opposite continent and mingled with the native varieties. They were chiefly imported from Holland, the cows of which country were the most celebrated of all others in the north of Europe for the abundance of their milk and the uses of the dairy. . . . . The Dutch breed was especially established in the district of Holderness, on the north side of the estuary of the Humber, whence it extended northward through the plains of Yorkshire; and the cattle of Holderness still retain the distinct traces of their Dutch origin, and were long regarded as the finest dairy cows of England. Further to the north, in the fertile district of the Tees, importations likewise took place of the cattle of the opposite countries, sometimes from Holland and sometimes by the way of Hamburg from Holstein, or the countries on the Elbe. Sir William St. Quinton, of Scampston, is said to have procured bulls and cows from Holland, for the purpose of breeding, previous to the middle of the last century, and, at a later period, Mr. Michael Dobinson, in the county of Durham, visited Holland for the purpose of selecting bulls of the Dutch breed. Other persons had resorted for their breeding cattle *to Holstein, whence the finest of the Dutch breed have themselves been derived.* Of the precise extent of these early importations we are imperfectly informed; but that they exercised a great influence on the native stock appears from this circumstance,—that the breed formed by the mixture became familiarly known as the *Dutch or Holstein* breed, under which name it extended northward through Northumberland and became naturalized in the south of Scotland. It was also known as the Teeswater, or simply the Shorthorned breed. . . . . . The breed communicates its characteristics readily to all others, and the first progeny, even with races the most dissimilar, is usually fine. The females retain in a considerable degree the properties of the *Holstein race*, in yielding a large quantity of milk, in which respect they greatly excel the longhorns, the Herefords, and the Devons. In the property of yielding milk, however, the new breed is inferior to the older and less cultivated one, showing that refinement in breeding and a greater tendency to produce fat are unfavorable to the secretion of milk. . . . . .

"The district of Holderness, it has been said, early obtained cows from Holland, and became distinguished beyond any other part of England for the excellence of its dairy stock. Many cows of the Holderness variety are yet to be found, but generally they have been more or less mixed with the Durham blood. The effect has been to improve the form but to impair their milking properties: nevertheless, the modern Holderness still stands in the front rank of dairy cows, and the great London dairies are chiefly supplied by them."

The same author, in writing of the dairy breed of Ayrshire, says, that "it is stated on competent authority that even so early as the middle of the last century the Earl of Marchmont had brought from his estates in Berwickshire, a bull and several cows which he had procured from the bishop of Durham, of the Teeswater breed, *then known by the name of the Holstein or Dutch breed*, and mention is made of other proprietors who brought to their parks *foreign cows, apparently of the same race.*"

He also refers to the Falkland breed of Fifeshire as "apparently derived from Holland, inheriting the milking properties of the Dutch races, but now nearly extinct in the pure state." It appears by his account of this breed that the county of Fife in Scotland has long been celebrated for the excellence of its cattle. When fat, they bring a higher price in the English market than any other breed, and the Fifeshire cows have also a high reputation as dairy stock.

These cattle are black, or black mixed with white, and are of a larger size than the cattle of the higher countries. The breed undoubtedly originated in the ancient domain of Falkland, situated in the lower part of the vale of Eden, in Fifeshire, and long the favorite retreat of the princes of the House of Stuart. It was the residence of James IV. of Scotland in 1502, when a treaty of perpetual peace was concluded between that monarch and Henry VII. of England, cemented by the marriage of James with Princess Margaret, eldest daughter of Henry, and a tradition has been handed down that the king received, with the dowry of his youthful queen, a present of 300 English cows, which were conveyed to the park of Falkland, whence their descendants spread into the neighboring country. "There is nothing," he says, "inconsistent with probability in this tradition, although the Falkland breed appears to be of foreign origin. It resembles the Black Dairy breed of the Low Countries, common in the dairies of Holland; and therefore, if brought from England, it must have been an imported race, though not the less likely on that account, to have been deemed a gift worthy a royal prince. . . . . The Flemings and Hollanders were even at this early period known for their cows, and it is altogether probable that some of these animals were brought to the royal park of Falkland as something that was curious and useful." And he makes honorable mention, also, of a new breed formed in Ireland, said to have been produced by selections of the best mountain cattle of that country, and expresses the existence of a doubt in his own mind whether the *Dutch blood* was not mixed with the native race in effecting the improvement.

Sanford Howard, in describing the Ayrshire breed of cattle, says, that "it is not improbable that the chief nucleus of the improved breed was the 'Dunlop stock,' so called, which appears to have been possessed by a distinguished family by the name of Dunlop, in the Cunningham district of Ayrshire, as early as 1780. *This stock was derived, at least in part, from animals imported from Holland.*" And Rawlin, writing of these cattle in 1794, says, that they "are allowed to be the best race for yielding milk in Great Britain or Ireland, not only for large quantities, but also for richness and quality."

Charles L. Flint says, "from the description given of these cattle, there is no doubt that they were the old Teeswater, or Dutch; *the foundation, also, according to the best authorities, of the modern improved Shorthorns.*"

L. F. Allen says: "In our history of the Shorthorns we have alluded to the probability that they were at a very early day, originally derived from the neighboring continent; and they may have descended from the same common ancestry to which *the improved breed of Holstein and Holland trace their lineage.* Their forms and general appearance, in all but color, indicate that they have sprung from a common source."

John C. Dillon, of the Massachusetts Agricultural College, writing in relation to the Holstein stock of the "Chenery importations," says: "They appear to me to possess, in an extraordinary degree, the qualities which distinguished *their descendants, the Shorthorns*, before that breed began to be raised for sale rather than practical usefulness, viz., large size, fine symmetry, vigorous constitutions, superior milking properties and a corresponding aptitude to take on flesh when dry."

And the "Deutsch Amerikanische Farmer Zeitung" has recently published an article upon Holstein cattle, in which it is stated that the Holstein race of marsh cattle have, for a long time, been known as affording excellent milch cows, and good, strong working oxen. In the first-mentioned property, it has, during the last hundred years, especially in North Holland, been brought to an *almost incredible degree of perfection*, since the farmers there direct their whole attention to the cows, whose products in butter and cheese enjoy a European reputation. *Yet, although this race of cattle has been most fully developed and attained to the greatest consequence in North Holland, the original stock was by no means bred in Holland, but in Holstein, whence it spread itself over the north of Germany and Holland, even to England, and contributed much to the improvement of the native stock of that country.*

The statements and opinions of the eminent authorities above quoted, are entitled to consideration; and it will be observed that the evidence adduced is sufficient to establish the truth of the statement before made, that a race of cattle of superior excellence existed in the Duchy of Holstein, Holland and adjacent provinces, at a period long anterior to the date of the commencement of attempts at the improvement of domesticated animals by the English farmers, and that the introduction of those cattle into the British Islands contributed greatly to the dairy qualities of the Shorthorns, Ayrshires and other British breeds.

In relation to the extraordinary superiority of the Holstein race of cattle existing on the European Continent during the present century, we have the concurrent testimony of eminent and observing agriculturists who have visited Holland and the adjacent provinces at various times during the last sixty years.

Professor Silliman, in his "Journal of Travels" in Holland, published in 1812, says: "*Innumerable multitudes of very fine cattle* were grazing upon the meadows; many of them were of a pure, milk-white color; others, nearly or quite black; but by far the greater number were marked by both these colors, intermixed in a very beautiful manner; and we found this fact to be general; for, wherever we went in Holland, the cattle were black or white, or striped and spotted with these colors. . . . We observed the cows in the meadows covered with blankets, to protect them from the dews." And the late Henry Colman, in his "European Agriculture," published in 1848, says: "*The Dutch cows have been a long time celebrated for their abundance of milk*, which does not surprise one in looking at the rich polders in which in summer they are fed, and where they are often seen covered with a cloth as a protection against both the dampness and the cold. . . . They are generally of a black-and-white color; in some cases they are milked three times a day. . . . They remain in pasture all summer, where they are milked; but in winter they make a part of the family, and, in truth, live in the common eating-room of the family, it being a part of the main house. . . . The cow stalls, while occupied by the cows, are frequently washed with water, . . . and over every stall is a cord suspended, by which the tail of the cow is tied when milked, to prevent her slapping the face of the milker or throwing any dirt into the pail. Indeed, the neatness of all their arrangements is perfect."

Another writer says the Holsteins, as seen on the polders of North Holland, or in Friesland, are unquestionably, as a race, *the largest yielders of milk in the world;* the average yield of a cow for the pasturage season of six months being twenty Dutch cans, or about twenty-seven quarts per day. And in a work published in 1858 by Charles L. Flint, Esq., Secretary of the Massachusetts Board of Agriculture, under the title of "Milch Cows and Dairy Farming," we have a treatise upon the dairy husbandry of Holland, translated from the German, giving valuable information in relation to the modern Holstein cattle, and affording abundant evidence of their superior excellence, especially in their adaptation to the cheese dairy. He says that the attention of farmers in Holland "is at the present time devoted especially to the dairy and the manufacture of butter and cheese. They support themselves, to a considerable extent, upon this branch of farming, and hence it is held in the highest respect, and carried to a greater degree of perfection, perhaps, than in any other part of the world. They are especially particular in the breeding, keeping and care of milch cows, as on them very much of their success depends.

"The principles on which they practise, in selecting a cow to breed from, are as follows: she should have, they say, considerable size, not less than four and a half or five feet girth, with a length of body corresponding; legs proportionately short; a finely formed head, with a forehead or face somewhat concave; clear, large, mild and sparkling eyes, yet with no expression of wildness; tolerably large and stout ears, standing out from the head; fine, well-curved horns; a rather short than long, thick, broad neck, well set against the chest and withers; the front part of the chest and the shoulders must be broad and fleshy; the low-hanging dewlap must be soft to the touch; the back and loins must be properly projected, somewhat broad, the bones not too sharp, but well covered with flesh; the animal should have long, curved ribs, which form a broad breast-bone; the body must be round and deep, but not sunken into a hanging belly; the rump must not be uneven; the hip-bones should not stand out too broad and spreading, but all the parts should be level and well filled up; a fine tail, set moderately high up, and tolerably long, but slender, with a thick, bushy tuft of hair at the end, hanging down below the hocks; the legs must be short and low, but strong in the bony structure; the knees broad, with flexible joints; the muscles and sinews must be firm and sound; the hoofs broad and flat, and the position of the legs natural, not too close and crowded; the hide, covered with fine glossy hair, must be soft and mellow to the touch, and set loose upon the body. A large, rather long, white and loose udder, extending well back, with four long teats, serves, also, as a characteristic mark of a good milch cow. Large and prominent milk veins must extend from the navel back to the udder; the belly of a good milch cow should not be too deep and hanging."

He further states that the cattle of North Holland are especially renowned for their dairy qualities, giving "not only a large quantity, but also a very good quality, so that a yield of sixteen to twenty-five cans at every milking is not rare."

A writer in the "New England Farmer," sketching farm life in Holland, gives a very interesting account of the dairy cows there. He describes the best cows as raised and kept by the cheese-makers of the Purmer, the Beemster and the Schermer in North Holland; there are also, he says, fine cows in Friesland, but nowhere so generally as in the Purmer and the Beemster. The cows "in the latter-named districts, and in all the better portions of Holland, give an average of twenty Dutch cans, equal to twenty-eight wine quarts per day, for the pasturage season of about six months, or all the people with whom I have conversed are mistaken." . . . "I have endeavored to learn what is the largest milking known, from a single cow in one day; but have not done so satisfactorily. At Elswout, a gentleman's place near Haarlem, the farmer, who was a very intelligent man of past fifty years, said he had never known of more than thirty cans (forty wine quarts); twenty cans he considered the average of good cows for the season.

"A large farmer of the Beemster knew a cow, many years ago, which gave thirty-four cans (forty-eight wine quarts) in a day. One farmer assured me he would not take more than twenty-four cans a day from a cow, seemingly thinking it was as much as she could afford to part with."

Prof. Geo. H. Cook, of the New Jersey Agricultural College, writing in 1871, says: "One of the first things that attract the attention of the traveller in Holland, is the great number of cattle." They are to be seen everywhere at pasture, and their decided colors of black and white make them conspicuous objects. The fame of the cows for dairy purposes led him to inquire into their peculiar excellences, and he visited two or three dairies, but got the most definite information at one in the Beemster. Wouter Sluis occupied the farm, and was thoroughly informed in his business. He judged of the qualities of his cows by the *size of the milk mirror*, the *yellowness of the skin*, the *abundance of scurf on it*, and the *clear definition of the black and white colors.*

"They were all carefully blanketed, and were constantly in the pasture; the weather was rainy and cool, but no shed was provided for them; and the conclusion was, that they would get no shelter until the winter stables were ready for them.

"Some of the cows, immediately after calving, give as much as twenty-five and even thirty-two quarts of milk per day.

"The wonder to a stranger is in the marvellous neatness of the cow stables. The cattle are turned out from them in the spring, and are kept in the pasture day and night until the grass fails. As soon as the stables are vacated, they are washed out clean, the floor is sanded, and in some cases tiles are laid, so that the stables are just as neat and clean as the dwelling-house, which is under the same roof, and is only separated from it by a partition and door. These stables are too neat for temporary use during the cold storms of spring or autumn; and as a substitute for this, the cows are almost uniformly covered with blankets or other cloths when in the pastures. When the cows are brought into the stables for the winter, they are kept with the same care that the best of horses receive with us,—the cattle being curried, and the stables being frequently and thoroughly cleaned and washed out with water."

He mentions one dairy of twenty-six cows in the Beemster, where the daily average of milk was 18.6 quarts per day for six months, and another at Haarlem, where the best cow gave 17.8 quarts per day for forty-four weeks;

and he further illustrates the extent of the dairy husbandry of Holland by stating that "there were in 1864, 1,333,887 cattle in Holland, of which 943,214, were cows; and these numbers are not so large as they have been in some other years; 32,000,000 pounds of butter and 61,000,000 pounds of cheese were exported from the country in 1864. The population of New York is about the same as that of Holland; the whole number of cattle of all sorts in that State, in 1870, was estimated at 702,000. The whole amount of butter exported from the United States, from June, 1869, to June, 1870, was 2,039,488 pounds; and of cheese for the same time, was 47,296,323 pounds."

This comparative statement, showing an amount of products in favor of Holland so remarkable, may be attributed, in some measure, to the peculiarities of the soil and climate of that country, but is, more probably, a result mainly of the careful and judicious selection and treatment of the cattle.

The importance of the dairy husbandry of North Holland and the adjacent dairy districts is also evinced by the uniform appearance of contentment, thrift and wealth exhibited by the population, which is exceeded by no agricultural community in the world; and this wealth, being a result of this particular branch of farming, it is easy to understand why the dairy farmer of this locality is a person of the highest respectability. And when it is considered that his success depends almost entirely upon his judgment in the selection, breeding and care of milch cows, it is easy to comprehend how it has come to pass that the dairy stock of Holland has attained to a degree of excellence surpassing that of any other race or breed; and in view of the inestimable value of the herds to the farmers of that country, it is not surprising that they "give their cows preference over everything else mortal. They are never overworked or underfed, as the wives and children sometimes are; they never lack blankets to keep them warm, nor shade to keep them cool; the warmest, best-built and best-kept portion of the house is set apart for their winter habitation; their food is prepared with strict attention to their tastes; attendants sleep in their apartments to see that no harm comes to them at night; milkers are regularly roused to their duties at three o'clock in the morning, and during the day a door is generally open from their stalls to the rooms inhabited by the biped members of the family."

The Holstein cattle were introduced into this country about the year 1625, by the "West India Company," and other importations were subsequently made into the State of New York by the early Dutch settlers there. At a later date,—about the year 1810,—the late Hon. Wm. Jarvis, the importer of Merino sheep, brought over a bull and two cows, which he placed upon his farm in Weathersfield, Vermont, where they were bred successfully for a time, and acquired a good local reputation; but not appreciating the value and importance of purity of blood in cattle, they were mixed with other breeds; and we are informed by one of his nearest surviving relatives that, several years previous to the death of Mr. Jarvis, which occurred in 1859, the pure blood of his importation had become extinct. [The writer would here state that in a paper upon Holstein cattle, published in the report of the Department of Agriculture for 1864, he was led to make the erroneous statement that some pure blood of the Jarvis stock was then in existence.] "Ten or fifteen years later," says Allen, "the late Mr. Herman Le Roy, of New York, imported some improved Dutch [Holstein] cattle into that city, and kept them on a farm in its vicinity. Some of them were, about the years 1827–8–9, sent to the farm of his son, the late Edward A. Le Roy, on the Genesee River, in that State. We saw them and their produce there in 1833. They were large, well-spread cattle, black and white in color, and remarkable for their uncommon yield of milk. . . . In the herds of both father and son, the *pure* breed was lost, as none but *grades* were found in the herds subsequent to the sales of the farms of those gentlemen a few years afterwards. It is to be regretted that the blood of these importations should have been so soon lost by a lack of interest in their propagation. They were of great value as dairy animals, as their qualities in that line were universally acknowledged where they were known."

Some of the small, belted or sheeted cattle, originally from the Swiss Canton of Appenzelle, have been imported from South Holland into the States of New York, Pennsylvania and New Jersey, where they have been found to be good dairy cattle, but deficient in size for the yoke or the shambles. These cattle are not to be confounded with the Holstein race, as they are a distinct breed or family, comparatively small in size, always reproducing its distinguishing feature of a broad white band extending from the withers almost to the hips, covering the sides of the barrel; the head, neck, fore-legs, shoulders, as well as the hinder parts of the body and legs, being black, or sometimes varied in shade.

It will be observed that although the cattle of the early importations above described were appreciated in the particular localities where they were introduced, yet breeders at that time, not realizing the value of *blood* and the importance of careful breeding, had crossed and intermixed them with other stock, until the breed, in its purity, had become extinct; and it is to the later importations, made by the writer, known as the "Chenery importations," that the Holstein stock mainly owes its reputation at the present time in this country, and (with the exception of recent importations, which will be alluded to hereafter), all the cattle in the United States recognized by the Holstein Breeders' Association as pure Holsteins, have sprung from said importations, and will be found duly recorded in the Holstein Herd Book.

The considerations which, in the first instance, led to these importations, were a confident belief in the decided superiority of the dairy cows of North Holland, in comparison with all other breeds, and also in their adaptation to the climate of New England, which, in its variableness, is strikingly similar to that of Holland, where alternate violent gales, dense fogs and extremes of cold and heat prevail,—the temperature ranging from 23° below to 102° above zero.

The first of these importations was made in 1852, and consisted of a single cow. The extraordinary good qualities possessed by that cow led, in 1857, to a further importation of a bull and two cows, and, in 1859, to four more cows. In consequence of a disease which occurred in 1859–60, these cattle, and all their full-blooded descendants, with the single exception of a young bull, were seized and destroyed under a law of the Commonwealth of Massachusetts; and in the autumn of 1861, another importation of a bull and four cows was made from North Holland, arriving at the port of Boston on the 6th of November, in good order and condition, after a voyage of thirty-six days. These animals, with their progeny, have been kept and propagated at the Highland Stock Farm in Belmont, Massachusetts, and, as before stated, formed the ground-work of the present Holstein stock of this country. The orginal animals were procured from amongst the best breeders in the vicinity of the Beemster and Purmerend, in the province of North Holland. In their selection, special attention was given to sanitary conditions, close consanguinity was scrupulously avoided, and animals of undoubted purity of blood, as well as those possessing to the full extent the peculiar characteristics ot the race, were obtained. All of which was substantiated by certificates of the burgomaster of Opperdoes, the mayor and magistrate of the community of Midwould, where they were obtained, the secretary of the Dutch Agricultural Society for the district Opmeer, and J. F. W. Korndorffer, first-class veterinarian of the kingdom of Holland. Experience, so far, has proved these cattle to be remarkably docile and quiet, both in stable and pasture; sure breeders and excellent nurses; easily kept, and as hardy as the native or "old red stock of New England"; and, moreover, the appearance of the young stock presents strong indications that the progeny will be superior in size to the imported animals. A description of some of the prominent animals of this herd, published in 1869, will, perhaps, convey to the reader a very correct idea of the prevailing characteristics of the Holstein cattle, as they exist at the present time on the best dairy farms of North Holland and the adjacent provinces.

The bull "Van Tromp," (see portrait), imported in the womb of "Texelaar," is now six years old, and his girth is 8 feet 5 inches; length, 9 feet 2 inches; height, 5 feet 2 inches; weight, 2,720 pounds; and the weight of the two-years-old bull, "Opperdoes 7th," is 1,597 pounds. The weight of the imported cow "Texelaar" (see portrait), is 1,560 pounds; "Lady Midwould" (see portrait), 1,620 pounds; the four-years-old heifer "Opperdoes 3d" (see portrait), 1,495 pounds; the three-years-old heifer "Texelaar 5th," 1,500; the two-years-old heifer "Texelaar 8th," 1,290 pounds; the yearling heifer "Zuider Zee 5th," 900 pounds; the bull-calf "Duke of Belmont," nine months old, 710 pounds, and the heifer calf "Midwould 8th," nine months old, 635 pounds; all raised in the ordinary way, without forcing, the young animals running in pasture from May until November.

At the present time (February, 1872), we give, as a specimen, raised on ordinary keep, the bull-calf "4th Earl of Middlesex," calved March 4th, 1871, weight, 862 pounds.

The color of these cattle is unique, being uniformly jet-black and snow-white, contrasted in the most beautiful and picturesque manner; their heads are well formed; eyes clear, large, mild and sparkling; horns fine, short and well curved, and in general answering the description given in preceding remarks.

An examination of this stock will, it is believed, afford conclusive evidence that while the English cattle-breeders have been laboring to perfect the beef-producing properties of the Shorthorn breed, the North Holland and Friesland farmers have been improving their dairy cows, until they have attained preëminence over all other breeds or races of cattle; and this is further and strikingly exemplified in the fact, that various writers upon dairy husbandry have, with remarkable unanimity, in furnishing illustrations of the best type of dairy cow, selected their models from the Holstein race. Professor Magne, of the Veterinary School at Alfort, in his little work entitled "How to Choose a Milk Cow," gives a specimen of this class which gave seventy English pints of milk. She is described as having a large pelvis, haunches widely parted, udder, veins and escutcheon largely developed. Cows of this class, he says, "when newly calved, and after several calvings, if fed with good, wholesome, moist and abundant food, well suited for the secretion of milk, are able to give as much as a pint of milk for every ten ounces which they consume of hay, or its equivalent in other food. They give milk for a very long period; the best never become dry, but continue to be milked, giving, up to the moment of calving, 17½, 21 or 26 pints of milk daily." Flint, in his "Milch Cows and Dairy Farming," gives the same figure, and states that she "was giving daily 22 quarts of milk a year after calving." And Allen, in his "American Cattle," gives an accurate portrait of the Holstein cow Lady Midwould, "showing a perfect *escutcheon*, developing, in the highest degree, the milk marks, by the escutcheon, of a largely-producing udder and its connections." She stands, he says, "*the model of a perfect milker*, with all the mammary veins and udder glands in the highest state of development."

The extraordinary milking quality of the Holstein is also well illustrated in the record of the imported cow Texelaar (see portrait). This cow was tested when only six years old, and therefore before having arrived at full maturity. She dropped a heifer calf on the 15th of May, which weighed, at birth, 101 pounds; and from the 26th of May to the 27th of July, a period of nine weeks, a careful and exact record of the milk produced by her was kept, showing a result of 4,018 lbs. 14 oz. Her largest yield in one day was 76 lbs. 5 oz. (35⅛ qts.), and in ten days she gave 744 lbs. 12 oz., or an average of $74\frac{47}{100}$ pounds per day. She continued to give a large flow of milk throughout the season, and was milked up to the 24th day of May following, and on the 25th dropped twin heifer-calves, which weighed, at birth, 155 pounds; and, notwithstanding the large quantity of milk produced, the quality is very superior, as shown by the statement of Prof. A. A. Hayes, Massachusetts State Assayer, who was employed to make chemical analysis of the milk of the four imported cows, and who stated that the "Texelaar's" milk "afforded, after long repose, $22\frac{73}{100}$ per cent. of cream in vessels specially adapted to measuring it." These cows, it has been said, are especially valuable in the cheese dairy; but, having no ready means of making a practical test in that respect, six days' milk of the cow "Texelaar" was set for cream, and when churned, it produced 17 lbs. 14 oz. of good butter,—nearly three pounds per day. In this connection it should be stated, that the "Texelaar" is not considered the largest milker in the herd. In the estimate of good judges, the cow "Lady Midwould" is her superior, although no favorable opportunity has offered for testing the latter—her calves having been dropped late in the season.

The report of the analysis above referred to, which was made at an unfavorable season (December), was as follows:—

"*Results of analysis of four samples of milk received from Winthrop W. Chenery, Esq.*

"The cans containing the evening milking of the cows reached me early on the following morning. Each can was sealed and accompanied by a certificate of Mr. George H. Nichols, superintendent of Mr. Chenery's farm, who had put the milk in the cans. On opening the cans immediately, the milk in each was found fresh and cool, and its delicate organization uninjured.

"All the samples at 60° F. were from one-half to one degree above the average of Orange County milk by lactometer."

"One thousand parts by volume afforded the following weights of constituents in samples:—

	No. 1. Texelaar.	No. 2. Lady Midwould.	No. 3. Zuider Zee.	No. 4. Maid of Opperdoes.
Water, (produced), . .	850 20	879.30	874.40	869 59
Caseine and Albumen, .	55.40	38.15	48 01	49 68
Sugar and salt, . . .	44.40	44.84	42 04	36.75
Pure butter, . . .	47.50	33 96	32 50	40 23
Phosphates, as bone phos, .	2.50	3.75	3 05	3.75
	1,000.00	1,000.00	1,000.00	1,000.00

"These milks, and especially No. 1, contain a larger quantity of albuminous matter than any samples which I have analyzed. This substance, found in all good milk, cannot be separated from the caseine so as to enable us to weigh it, and I have been compelled to include it with the caseine found. The albuminous substance is not only highly nutritious as a diet, but in the cases of the samples it confers a singular constitution on the milk, considered as an organized secretion. It divides the pure fatty part of the milk in a way to prevent it from rising in the form of cream copiously, and holds a part of it in what would be the skimmed milk, rendering it necessary, in order to obtain all the butter, that the *milk*, instead of the *cream*, should be churned. But its office has a more important connection with the actual nutritive power of the milk, which it increases greatly in two ways: 1st. It is itself a highly nitrogenized product. 2d. It is in these milks so balanced in connection with the butter as to be easily assimilated and digested without coagulation. These are valuable properties in their relation to the rearing of the young of the human or animal species, and I should expect to find these milks to possess fattening properties to an extraordinary degree, as indicated by the analysis.

"The subject has interested me much during the progress of the experiment, and a further collection of facts will probably prove that a substitution of albumen for caseine may often occur in the constitution of milk, and this change may be closely connected with nutrition. . . . .

"When these milks were inspected under the microscope, the arrangement of the fat globules and the general organized structure of the mass of the milk, seemed to result from a balance maintained by the albuminous part of the constituents. After the milk had become sour, the whey did not drain from it as usual, nor was the albumen coagulated after many days,—properties not usually observed in milk."

These milks were also analyzed by Dr. C. T. Jackson, of Boston, Mass., with similar results. And, for the purpose of comparison with European analyses of milk, he cites, in his report, analyses of cows' milk by Boussingault and Poggiale, and says, "by comparing the analysis of your cows' milk with theirs, you will perceive that your samples are richer than those analyzed by Boussingault and Poggiale, and you will also observe that the milk of your cows is *specially adapted to making cheese*, since it is unusually rich in caseine, which is the basis of cheese."

The results of the chemical analysis, above stated, it will be observed, indicate the milk of the Holstein cows to possess fattening qualities and nutritive power to an extraordinary degree; and, inasmuch as it is peculiarly rich in the constituents which go to form the bones, muscle and fibrous tissue, and also in that property which serves to give heat to the system in its combustion in the circulation, it is particularly adapted to the rearing of the young of the human, as well as of the animal species. And, moreover, as the analysis shows (what Dr. Jackson states and experience has confirmed), this milk to be specially adapted to making cheese, and as cheese-making has become one of the most important agricultural interests in the country, and cheese factories are rapidly multiplying in various localities, it may be well to consider the comparative value of these cows for that specialty. In this view, reference is made to a statement translated from a paper received at the Department of Agriculture, Washington, from the Royal Academy of Agriculture, Prussia, written by Dr. Rhode, of the Agricultural Academy at Eldena, from which valuable information is obtained in relation to the extraordinary milking qualities of *Holstein cows.* The herd described consisted of 36 cows, and in 1865 a record of nine superior *pure-bred cows* was kept, showing a yield respcctively of 4,960 quarts, 4,710 quarts, 4,620 quarts, 4,490 quarts, 4,365 quarts, 4,800 quarts, 5,016 quarts, 5,009 quarts, and 4,900 quarts, or an annual average product of more than 4,700 quarts of milk from each cow. In the translation it is not stated whether the quarts were beer or wine measure, but, supposing it to be the smallest measure and the lowest estimate of two and one-sixth pounds to the quart, the annual yield of the nine cows will be found to be 92,885 pounds of milk, or if made into cheese, allowing nine and one-half pounds of milk to one pound of cheese, as made at the New England and New York cheese factories, we have the astonishing result of 9,777.36 pounds, or an average of over 1,086 pounds of cheese annually from each cow, while the maximum yield of the best cow would be 1,144 pounds.

The above statement, it may be said, in passing, should furnish "food for thought" to the cheese dairymen of this country, who have seemingly been content heretofore with an average annual yield of about 300 pounds of cheese per cow, as shown by the various cheese factory reports. And inasmuch as experience has abundantly proved that the progeny of common or grade cows, when bred to Holstein bulls, uniformly inherit the peculiar characteristics of the Holstein race, may it not be a question of the highest importance for the consideration of dairymen, especially in localities where cheese-making is the leading interest, whether the introduction and use of Holstein bulls would not very materially raise the average products per cow, and enhance accordingly the profits of their business.

With regard to the grazing or fattening qualities of the "Chenery importation" of Holstein cattle, but little can be stated as the result of experience. All the males having been kept for bulls and the females for breeding cows, no full-blood animals have been fed for the shambles. It may, however, be stated that the young animals grow with astonishing rapidity and mature early; and that the milch cows, when not producing a large quantity of milk, show a remarkable tendency to take on flesh. These indications, together with the considerations of their uniformly large size, great muscular development, quiet disposition and early maturity, go far to settle the question of their value as meat-producing animals.

For working oxen, these cattle present material of the first order. Their symmetry of form, beautiful proportions and immense power, great intelligence and tractability, hardiness and remarkable capability of withstanding extremes of heat and cold, all conduce to place them in the front rank in the department of labor.

Grade Holstein steers have been fed and slaughtered with the most satisfactory results, and grade oxen have been raised attaining a live weight of 4,600 pounds, at five years of age, and having no superiors as workers; and this leads me to recur again to the extraordinary value of the Holsteins as a means of improvement by crossing them upon the native, inferior breeds, or mixed cattle, and to state that the well-known and acknowledged principle in breeding, viz.: that *blood* exerts an influence exactly in proportion to the antiquity of a race or breed, has been verified by the experience with these cattle at the Highland Stock Farm. Hundreds of cows, of almost every breed known in this country, have been bred to the Holstein bulls there, and it has been remarked that the calves of these crosses invariably resemble the sire, in color, shape and general appearance, all the half-bred progeny being exceedingly thrifty, the steers making very rapid growth, and the heifers uniformly inheriting the milking quality of the Holstein race.

The German editor of the "Deutsch Amerikanische Farmer Zeitung," in the article, to which we have before referred, in relation to these cattle, says that the forms of animals of the Holstein race are generally extended and angular, and the bones, notwithstanding their comparatively large bodies, only moderately heavy. The cows show unmistakable marks, indicating extraordinary abundance of milk, while the oxen, through their great size, are not only in condition to perform much and heavy work, but have also gained a great reputation when stalled and fatted as cattle for slaughter. Their color is for the most part spotted black and white; sometimes almost wholly white, and sometimes more black. Hide and hair fine and soft, the head long and narrow, with rather pointed muzzles. The horns are short, generally strongly curved forward. The neck moderately long, the breast not too broad, without prominent dewlap. The arches of the ribs not very broad, and not barrel-shaped. The belly long extended, reaching both down and back. The rump broad, moderately short, falling sharply behind, prominent haunch bones. The tail deep set on; moderately high on the legs; the hind-legs giving room for a large udder.

Although the value of this race of cattle was early known in this country, and examples brought here more than two hundred years ago, and others again about the twentieth year of this century, possessing the general characteristics above described, yet they could not gain ground against the almost universal predilection for the English breeds until Mr. Winthrop W. Chenery, of Massachusetts, undertook large importations, and having presented the stock in its purity, he has succeeded in raising cattle of so excellent a quality as to come in competition for public favor with the so much admired Shorthorns. The cows give a large quantity of milk, especially rich in caseine, while the oxen, whose performances in regard to their working capacity are extremely important, easily become fat on proper food, and supply much flesh of an excellent quality for the shambles. They have the advantage over the Shorthorns in not being so easily affected by our changing climate, and probably in the States of the North and North-West will thrive better than the shorthorns, which are on the whole somewhat pampered."

The favorable opinion entertained by the writer in regard to the merits of the Holstein cattle is held in common with all the breeders of the stock in this country, and it is exceedingly gratifying to be able to present the following extracts from the many commendatory letters received upon the subject from practical breeders in various States in the Union.

Mr. D. E. Brower, a member of the Doylestown Agricultural and Mechanics' Institute, of Doylestown, Penn., writes in relation to the Holstein bull, "Earl of Middlesex," purchased by the Society for the improvement of stock, thus: "Two years have now past since the Earl was brought into our county. For the year ending Feb. 9th, 1871, forty-seven cows were brought to the bull, and about forty of these dropped calves. For the year ending Feb. 9th, 1872, seventy-six cows were served; and, from what is now known, more than sixty are with calf. So that the Society will nearly pay for the first cost and expenses of keeping the bull in two years' time."

"From all the indications we have, the stock appears to be *just what is needed for our dairymen in Eastern Pennsylvania.*"

"The Earl of Middlesex is a very fine specimen of his class, and should we meet with the same encouragement to the end, which has thus far been realized, the Doylestown Society will have at least done one great thing for Bucks County.

"It occurs to me that we have no breed of imported cattle that can compare with the Holstein for constitutional vigor, and consequently for profit. If our people are not mistaken, this stock will make more meat and milk out of a certain quantity of food than any other."

Capt. Wm. Steckel, of Doylestown, says: the Holsteins "created quite an excitement at our fair; people generally are agreeably disappointed when they see them. They expect to see large, rough cattle; but instead of that they find them large, smooth and handsome. The bull is fine, and doing well, grows like a weed in rich soil; he has gained five hundred pounds in seven months, fed on hay, green corn, and grass; he is a splendid animal in all respects, very neat, and the touch very fine; skin yellow and soft; in fact his coat cannot be equalled, and he is getting very broad and square. He is withal, exceedingly gentle and easy to manage. I have seen several of his calves and they are all beauties; of good size, square-hipped, with broad shoulders, good coats, bright eyes, and stand square on their legs. One bull-calf weighed one hundred and sixty pounds when a few days old. Altogether, the calves are making quite an excitement among the farmers. The half-bloods come very fine, and several half-blood bull-calves have been sold at $50 each; there has been but one half-blood heifer sold in the county, and I sold her for $100 when four months old. *The farmers will not sell the heifers got by the Holstein bull.*

"My pure-bred Holstein heifer, 'Texelaar 9th,' calved April 11th, a heifer calf, in color like the dam; it is very fine in skin and hair, and shows good breeding in all points. I measured the heifer's udder the day before she calved. It measured 45 inches round, and from flank to flank, between the teats, 27 inches. As soon as the milk was good, we saved what the calf could not take, and when creamed and churned made 5½ pounds of splendid butter in seven days, on hay alone." Subsequently, he writes: "'Texelaar 9th' is large, weighing about 1,400 pounds, and has milked beyond my expectations. She gave, last May, after calving, 52½ pounds of milk a day, which is about 26 quarts, and one week's milk churned 12¾ pounds of superior butter; nine months after calving she gave 12 quarts of milk per day, and this with her first calf. She milks well yet, and I think will continue to do so all winter. She comes in again in May."

Mr. O. F. Jones, of Wooster, Ohio, writes, in relation to the Holstein stock, as follows: "In my judgment, there never has at any time been introduced into our county any cattle that have so completely commanded the attention of our people as the Holstein stock; and, as an evidence of the high favor in which the public hold this breed of cattle, I would say that the Holstein bull which I purchased of you has been patronized to the full extent of his ability since his arrival here. He has grown to be large and stately, but withal remains as gentle as a cow. I have never found it necessary to put a ring in his nose. I regard this breed of cattle as well adapted to our seasons, climate and pasturage. Their size and thrift, on ordinary keep, entitle them to a high rank amongst our farmers as a profitable breed for the shambles. I have a grade Holstein heifer which, for a period of about two months, subsequent to the 6th day of last May, at which time she dropped her first calf, gave a very fine flow of milk, averaging a little more than six gallons per diem. Her calf weighed, at five months, 400 pounds, having been kept in stable all the time." "There are quite a goodly number of the half-bred Holstein heifers here, *not one of which can be purchased at any reasonable figures*, on account of their very superior milking and butter qualities. This variety of cattle has fully met the expectations of their most sanguine friends, and the reputation of the stock is firmly established at this point. Our Holstein stock, including the old bull, is doing remarkably well, and all that you have ever written or spoken to me in favor of the cattle has been verified."

Mr. Lucian D. Trow, of the firm of Orin Trow & Son, eminent dairy farmers, of Hardwick, Worcester County, Mass., says: "It is with pleasure that I give you some facts and opinions relative to the Holstein cattle, based on our experience with them. Their renowned milking qualities had years ago become known to us through various writers on dairy husbandry, and we had a strong desire to test their merits by actual experiment in this well-known dairy region, and ascertain how far they were adapted to meet the demands of this particular section. We had a good herd of grade Shorthorns; and, regarding a cross with the Holstein a desirable one, we came to the trial honestly, without any bias or prejudice, and obtained the thoroughbred Holstein bull, 'Midwould 7th,' calved in May, 1866. Our first year's breeding was a total failure, owing to abortion in our herd. We were, however, so favorably impressed with the extraordinary growth and development of the bull that we resolved to persevere; and, having fattened our whole herd and turned them over to the butcher, we replenished our stock by new purchases, and have since been successful in our breeding. We reared last season six heifer calves, which are marvels for size, beauty, and promise, as clean in all their points as the best-bred Shorthorns; and showing the most remarkable marks for milk we have seen developed in any calves of their age. They were all taken from their dams at an early age, and fed about equal parts of milk and water until six weeks old, and then fed with whey as it came from the cheese factory. *We never raised six calves so uniformly superior, and none ever at less cost.* This spring we had the pleasure of hearing one of the best dairymen and Shorthorn breeders of Worcester West, after a close examination of these grade Holsteins, acknowledge that he had not seen their equals of *any* breed elsewhere in the county.

"We have now a bull-calf, raised on ordinary keeping, without any attempt at forcing a rapid growth, which at six months old weighed 475 pounds, and is now, at eight months, estimated to weigh 700 pounds, and he is absolutely perfect in shape.

"A rumor had found currency here, fostered by parties hostile to this race and interested in other breeds of cattle, to the effect that both the Holsteins and their grades were unprofitable for veal. Nothing could be more erroneous. During our breeding of this stock we have sent many calves to market, and have uniformly obtained the highest prices. We sent a calf last August, twelve weeks old; the dressed weight of carcase being 238 pounds, and pronounced by the butcher *the best calf he had ever killed*, and his sales amount to over 1,000 veal calves per year. Farmers who use grade bulls, the get of the Holstein bull, 'Midwould 7th,' say that their veal calves by these bulls bring nearly double the money (by reason of their large size and excellent fattening qualities) that they were accustomed to obtain for calves from the use of common bulls bred to the same cows. These are facts which can be substantiated, and our personal experience and observation lead us to believe that there are *no better veal calves in the world.*

"Very few of the grade Holstein heifers are yet in milk. One of our neighbors, however, has one which dropped her calf before she was two years old, and is now, several months after calving, giving 36 pounds of milk per day. Her owner says she gave a much larger quantity earlier in the season; and here we will say, that an average of 25 pounds per cow in a herd is considered a good one in this vicinity; comparatively few reach 30 pounds.

"As to the Holstein bull, 'Midwould 7th,' we have no hesitation in pronouncing him one of the best stock animals in New England. He made the best growth of any thing we ever had in our stable, and that on ordinary keeping. He never got a calf that was not a model for growth, shape and vigor; indeed, we have never known a bull that reproduced his like, in form, color, &c., with such unerring fidelity. Could he have been generally patronized by the farmers of this town for a period of five years, we believe it is a low estimate to say that he would have been worth as many thousands of dollars in the improvement of our herds. Now that he has gone, and doubts and prejudice are giving way to plain facts, which cannot be controverted, many others are already entertaining the same opinions as ourselves.

"In instituting a comparison of the Holsteins with other breeds, such as Shorthorns, Jersey, Devonshire, Ayrshire, &c., it is not necessary to attempt to detract from the peculiar merits of any of these different breeds. The Holsteins seem to us to rival the Shorthorns more properly than either of the other breeds enumerated, as to growth, size, beef and milk qualities; and experience has convinced us, that either thoroughbred or grade Holsteins will outgrow either thoroughbred or grade Shorthorns on our common keeping; and as to hardiness and vigor, the Holsteins have no superiors.

"After an experience of upwards of forty years in stock-breeding, which we trust has not been without its lessons, we are compelled to express the conviction that the *Holsteins are the best cattle for all purposes* we have ever known. Had our convictions been otherwise we should not have purchased, as we did last autumn, from your herd, the young bull, '3d Duke of Belmont,' for the further improvement of our stock."

S. B. Emerson, of Mountain View, Santa Clara County, California, says, in relation to the Holstein cattle: "I am satisfied that they are just what we want;" and the Corresponding Secretary of the California State Board of Agriculture writes thus: "We are glad to learn to-day, through S. B. Emerson, Santa Clara County, that some of that valuable stock, the 'Holstein,' has found its way into our State . . . . We shall give this breed the same consideration in reference to premiums at our fairs as other full-blood cattle."

Messrs. W. E. & B. Simpson, of Cambridgeport, Mass., give the following statement with regard to the production of milk by a half Holstein cow, sired by the imported bull, "Dutchman."

"She gave 6,390 quarts in a year, and her largest yield in one day was 34 quarts. She has never been forced by high feeding, having been fed upon salt hay and shorts, and sometimes, but not often, one quart of meal per day. We milked 30 quarts a day from her for about three months. Her last calf was seven months old on the 13th inst. [June, 1869]; and she is now giving, on grass feed alone, from 20 to 24 quarts of milk per day. We sell the milk of our cow at nine cents per quart. You will, therefore, see that the gross income from her for a year, exclusive of the value of her calf, was $575.10."

Asa G. Sheldon, of Wilmington, Mass., says: "After forty years' experience, and trying almost every breed of cattle that came within my knowledge, I think the Dutch [Holstein] breed imported by Mr. Chenery, excel all other breeds in three points: first, for early beef; second, for working oxen; third, for large quantities of milk."

Henry Rhodes, of Oneida County, N. Y., writes: "As regards your Holstein cattle, be assured I have a very high opinion of them for dairy purposes. I am very glad they are becoming more extensively known by their introduction to various other localities. . . . . Allow me to say, that, in my opinion, the publication of a Holstein Herd Book is a step in the right direction, for I am most certain that they are not sufficiently known through the dairy sections of this State to distinguish them from other importations of a different breed of Dutch cattle."

Z. E. Jameson, says, in the "Country Gentleman," January 14th, 1869: "At the present time there seems to be an increase of interest among cattle breeders in all sections of the country, in the introduction of full-blood stock. All kinds find admirers and purchasers; but among late importations, there seems to have been none that so fully commends itself to the judgment of farmers, as that of the Dutch [Holstein] cattle, by W. W. Chenery, Boston, Mass. Their excellences were first brought to public notice in the Report of the Department of Agriculture, in 1864. Then, at the fairs in the New England States, they have attracted much attention on account of their color, which is black and white, and size, which rivals that of the well-fed Shorthorns, and the appearance that indicates they are as good milkers as their owners represent them to be. . . . . At present there is no Dutch [Holstein] Herd Book; but it is evident that breeders should give attention to this subject now, *as all pure-bred animals have been imported or bred since* 1861, and now there would be little danger of fraud or mistake in giving the history of individuals of this breed."

The above suggestions were wise and considerate; but foreseeing the importance of preserving correct records, and anticipating the possibility of attempts at fraud, and, also, the probability of mistakes by breeders in furnishing pedigrees, the writer had, from the date of the first importation, preserved the history of every individual of this stock, and is, therefore, able to say, without any qualification, that the pedigree of every animal in the accompanying herd record can be substantiated, and it is also gratifying to know that said precaution and painstaking have frustrated some attempts to obtain record for grades as pure bloods.

Specimens of these Holstein cattle have been exhibited at some of the agricultural fairs in New England, and have received the highest prizes, medals and diplomas of the various societies; and the following extracts will show that they have also commanded public admiration as well as the appreciative commendation of practical breeders.

[From the Transactions of the New England Agricultural Society for 1864.]

"The committee to whom was referred the consideration of Mr. Chenery's herd of cattle of the Dutch [Holstein] breed, desire to express the very great gratification they have experienced in surveying the fine proportions of these noble animals, and of perusing the statements of their milk-producing capacities. Their presence here has been a marked feature of the exhibition, *and the committee cannot doubt that their importation and diffusion is to have in the future a most beneficial influence upon the stock of the country.* Although, from their great size and height from the ground, they seem little adapted to graze our short New England pastures, *there can be no doubt of their admirable adaptation to the fertile lands of the West, and such other parts of the country as afford abundance of pasture and winter forage.* It is claimed for the milk of these animals, and *chemical analysis proves it*, that it is *wonderfully rich in the constituents of cheese;* and if it should be demonstrated that they will thrive in such localities as will profitably support the Shorthorn, and furnish a greater weight of beef and a larger amount of a commodity so merchantable as cheese, and so easily transported, Mr. Chenery may well claim to have been one of the country's benefactors by his liberal outlay of capital in securing such superb animals for his breeding stock, and for the care and skill he has bestowed in their breeding."—*Report of Committee upon Dutch [Holstein] Cattle.*

[From the Transactions of the Middlesex Agricultural Society for the year 1864.]

"The herd entered for exhibition solely, was the famous Dutch [Holstein] herd belonging to Winthrop W. Chenery, Esq., of Belmont, president of the society, which *having taken the first premium last year*, could not, of course, be entered as a competing herd this year.

"It is needless for the committee to say, that Mr. Chenery's herd *would constitute an object of prime attraction at any agricultural fair in the land*, and they were obliged to him for again placing his magnificent cattle on exhibition the present year. They are cattle worthy the land of their nativity,—the land of the dairy *par excellence.* To the statement made by Mr.

Chenery last year, and published in the Transactions of the society, the committee, of course, can add nothing in the way of information. The herd, whenever and wherever seen, speaks for itself."—*Extract from the Report on Herds of Cattle.*

[From the Report of Middlesex North Agricultural Society for the year 1865]

"The heifer calf I offer for premium is three and a half months old, 'grade Dutch' [Holstein]. She is the first calf of a three-year-old heifer. The udder of her mother, one week before she dropped her calf, was so full, and the weather so hot that I thought it best to milk her. She filled the largest size Shaker pail full of milk. This milking was continued once a day for four days, her milk increasing, and on the fourth day she gave two quarts more than the pail full. The *fifth, sixth and seventh days she was milked twice each day, giving eighteen quarts per day, by actual measurement, and on the morning of the eighth day dropped her calf.* She continued to give a larger mess of milk than her calf would take. *On the day the calf was three weeks old, she gave eleven quarts of milk by measure more than her calf would suck.* I think she was the greatest heifer for milk I ever saw. I think this is the best story I can tell in favor of the Dutch [Holstein] breed of cows."—*Extract from Statement of Asa G. Sheldon.*

[From the Transactions of the Hampshire Agricultural Society for the year 1871.]

"In striking contrast with the little Brittanies, are the Holsteins, a breed of large black-and-white cattle, imported by Winthrop W. Chenery, Esq., of Belmont, and since by other parties. The inhabitants of North Holland, and the adjacent provinces have long been famous for the possession of a superior race of shorthorned milking cattle, which, it is claimed, originated in the Duchy of Holstein; and Dutch energy and perseverance have been successfully exerted to fix and increase their valuable properties. These properties are precisely what we should expect as the result of judgment and experience in breeding, and care and skill in managing, combined with the most favorable conditions of soil and climate; viz., large size, symmetry, second only to the best improved Shorthorns, an extraordinary capacity for milk, and a proportionate aptitude to lay on flesh when dry. The milk of the Holsteins is rich in casein, albumen, and sugar; it also contains a full average percentage of butter; but it is lacking in the tendency to separation which, in twelve hours, converts Jersey milk into cream and blue water. . . . . We feel sure the majestic Holsteins will find a broad field and a warm welcome in New England Agriculture."—*Extract from Report on Milch Cows and Heifers.*

[From Lewis F. Allen's new work on American Cattle.]

"There is a class or breed of cattle now existing in North Holland which have been greatly improved within the last century. That is eminently a dairy country, and the cows of the farmers and dairymen there receive a care and attention beyond any other domestic animal used in the agriculture of the people. . . . .

"Of the time at which any very considerable *improvement* was attempted in the Holstein cattle, we have no definite knowledge. It must have been more than a century, perhaps two or three centuries, ago; as it is only by a continuous and fixed system of breeding, for a long time, that the *undeviating*, constitutional characteristics of any breed of cattle can become so established as to transmit them with entire certainty to their progeny. These characteristics the present improved Holstein cattle *do* obviously possess in a sufficient degree to class them as a breed by themselves; and, as such, we shall treat them.

"Their *surpassing* excellence appears to be in their milking qualities, coupled with large size, and a compact, massive frame, capable of making good beef; and, in the oxen, strong laboring animals. They are almost invariably black-and-white in color, spotted, pied, or mottled in picturesque inequalities of proportion over the body. The horn is short, and the hair is short, fine and silky. *The lacteal formations in the cows are wonderful*, thus giving them their preëminence in the dairy. Our illustrations will show these prominent characteristics so plainly that further description is unnecessary. . . . .

"For the dairy, the qualities of the Holsteins must be acknowledged as remarkable. The Shorthorns, as in many instances of trial, have hitherto acknowledged no superior; yet they have now, in these new strangers to our soil, to say the least, found most formidable competitors, and an opportunity is here offered, by those who cultivate them for the dairy, to test their long acknowledged good qualities by comparison. The Holsteins have been long bred and cultivated with a view to develop their lacteal production to the utmost; and that they are quick feeders, and physiologically constituted to turn their food readily to milk, must be evident. . . . .

"We are gratified that this valuable importation has been made by its public-spirited owner, for the benefit of our cattle and dairy interests, and trust that their merits will spread far and wide, beyond the limited territory where they have, in such brief time, been so thoroughly tested. The grade heifers, by the Holstein bull, on other cows of different breeds, are said to inherit much of the good milking qualities of the Dutch blood.

"As a beef animal, their merits have been, as yet, but partially tried in the half-breeds, or grades from the Holstein bull, on the natives, or other cows of different breeds. So far, however, they are claimed to be satisfactory, ... We have personally seen and examined several of the animals of this herd, and from those specimens—to which the portraits do no more than justice—we should pronounce them good grazing and feeding cattle, in addition to their preëminence for the dairy.

"As a working ox, they will probably rank with other heavy cattle of like quality—better in their grades with the lighter and more active breeds, no doubt, than in the thoroughbreds—as with the Shorthorn crosses. They are entitled to a fair trial; and, in the hands of proper parties, their entire merits cannot fail to be thoroughly and advantageously developed. We consider Mr. Chenery's importation a decided acquisition to the cattle interests of our country, and trust that they will become widely known and distributed."

[From Willard's Practical Dairy Husbandry.]

"The Holstein or Dutch cattle, of North Holland, are perhaps the most noted for the dairy of any originating on the Continent of Europe. Holland has long been a dairy country, and the farmers there have given more care and attention to their cows than to any other domestic animal. The breed is of large size, and of a compact, massive frame, capable of making good beef. Their color is black-and-white, spotted or mottled in picturesque inequalities of proportion on the body. The horn is short, and the hair short, fine and silky. The Holsteins have been long bred and cultivated with a view to develop their lacteal production, and their milking qualities are truly wonderful. Some have expressed doubts, whether cows of so large a size, weighing from thirteen to fifteen hundred pounds, could be made profitable on much of our dairy lands. Doubtless, on poor, thin soils their introduction might not prove advisable; but when there is an abundance of food, the case would be different. Holland cattle have as yet received but little attention from American dairymen; but, if the European accounts given of them are to be credited, there is reason to believe they would prove a success, on our deep, rich soils or most productive pasture lands. . . . . The Holsteins cows have a reputation of being specially adapted to cheese dairying, and it is for this purpose, doubtless, they should be employed."

[Extracts from a Letter of Hon. T. S. Lang, in relation to the Holstein Cattle.]

"Dear Sir:—I must herein claim your indulgence for not returning an early answer to your inquiries concerning my opinion of the Holstein cattle ('as I see them').

"I well know you need no endorsement of mine to convince you of the value of this stock to the breeding interests of this country.

"When I last saw you, before leaving for Europe, in 1866, I was unwilling to give an opinion formed upon so slight acquaintance as I then had with the stock, but observation since my return and while abroad, makes me feel cheerful to encourage you all I can in the introduction of this breed of cattle. . . . .

"I feel sure that I wish to speak guardedly upon a subject of this kind, and I have endeavored to compare this stock with others with which I have been more or less acquainted.

"My father, J. D. Lang, more than twenty years ago, while I lived at home, commenced breeding Shorthorned stock of the Greene stock of Albany, keeping sometimes 40 head. This stock was unfortunate, on account of a constitutional defect, which in the end run the stock down very low. They were not satisfactory as a dairy stock.

"Feeling that the trial of Shorthorned stock by my father was unsatisfactory, on account of the peculiar hereditary difficulties attached to that special family, I bought six cows and a bull of Samuel Thorne's splendid stock. They were beautiful animals; some of them had gained a first-class reputation in this country, and in England before they were imported. The progeny of this stock was the finest I have ever seen in Maine of Shorthorned stock. While I had this stock I saw your Dutch [Holstein] cattle at Springfield; and, being struck with their peculiar vigor and milking qualifications, so different from the cows which I then had, I desired to buy them as an experiment in crossing with the Shorthorned stock. I had no idea of making a specialty of them, or any other stock, without comparison with other breeds, that I might be able to see for myself what appeared to be best adapted to the wants of Maine.

"I do not wish to convey to you the idea that I believe Dutch [Holstein] stock, or any other breed, are suitable to every locality among us. But were I situated upon clay loam soil, or good grass land, I should decidedly choose Dutch [Holstein] and their cross upon Short-horned cows as the *sine qua non* of my wishes. I do not wish to be understood as encouraging any one to breed them for the *special* purpose of butter-making; but as a stock uniting a hardy, vigorous constitution—which it proves by developing the most remarkable growth of bone and muscle—with a power to assimilate thereto a more varied and cheap class of food than any other stock that I have ever met. One point which struck me as peculiarly desirable, was the distribution of fat through the animal, rather than in lumps or deposits upon the surface.

"The half-bloods of this stock are remarkable for their thrift. A few days since, I met a butcher who does a large business here, and sends much meat to market; he said it was unfortunate that I had allowed the Dutch [Holstein] bull to go out of Maine, and remarked that he paid from four to five dollars per head more for calves four weeks old from this bull than any other.

"I am unwilling to have you think that I underrate Shorthorned stock; on the contrary, I admire them; and, in breeding a mixed stock, I should certainly aim to select cows of that breed, whenever they could be found prime milkers, or feeders of their progeny. So far as I can judge, the cross of Dutch [Holstein] bulls on Shorthorn cows is admirable. . . . .

"My conviction is that, to institute an experiment of ten or more cows of Dutch [Holstein] and Shorthorn stock under the same feed and circumstances, as far as practicable, and weigh the progeny at six months or two years old, it would be found that the Dutch [Holstein] stock would cost the least per pound. My experiments with the half-bloods were very gratifying, and outstripped all others with same feed. These experiments were made with some care. I repeatedly received offers of $255 to $250 per yoke for eighteen months' steers, *which never ate grain.* I sold four to your neighbor, Mr. Munroe, which girted, at eighteen months, 6 feet 7 inches average, and one pair averaged 6 feet 8 inches.

"As to the working qualities of Dutch [Holstein] cattle, or their crosses, I am unable to give you any idea from personal knowledge. I questioned the herdsmen in two or three estates in Belgium, where these cattle were kept for cheese-making, and they assured me that they excelled in this particular; one of them, pointing to the intelligent head and eye, and strong, straight, active limbs, saying, 'Do you doubt it?'"

In another letter, the same gentleman says: "I assure you that I would not take the gift of the best of Shorthorned stock for my purposes, for pounds of beef in the shortest time, if Dutch [Holstein], like the bull I had, could be bought. Some of his calves this year are wonders (yearlings). I am aware that this is strong; but would you buy anything short of the best seed to sow or plant, especially in a business which has a period of several years to develop results? I congratulate you upon your stock."

The recent importations of Holstein cattle, to which reference has been made, are as follows:—Mr Gerrit S. Miller imported from Friesland, in October, 1869, a bull and three cows, and is now breeding them at his farm in Peterboro', Madison County, N. Y. The writer has not seen this importation; but, judging from the representations of Mr. Miller, and the following description, published in the "Utica Herald," it is evident that the animals possess, to the full extent, the characteristics of the Holstein race, as described in these pages.

"They are black-and-white, the spots being large and clearly defined, and quite thin in flesh, having the appearance of cows that have had poor feed, and run to milk instead of flesh. Their frames are very large and capacious, having the square build which we are wont to associate with the Shorthorns. Their necks and heads are somewhat thicker and heavier; their muzzles are larger, and their horns are not as fine. But the milk-mirror, the milk veins and the udder, all indicate extraordinary milking qualities. The skin is about as yellow as the dandelion, and would show them to be excellent for butter, if the popular sign is infallable; but stress is not laid on their butter-making qualities, although Mr. Miller told us that from one of his cows he had made 13 pounds of butter in one week. Their extraordinary flow of milk, and its richness in caseine, would seem to make them very valuable for cheese-making. In one week this season, the 'Crown Princess,' six years old, averaged 70 pounds of milk a day, and her largest day's yield was 74 pounds. We were furnished the following statistics of the yield of milk of three cows, during this season, the milk being sent to the cheese factory. They were fed four quarts of meal each a day, until the 5th of June, after which time, until the 5th of September, they had only what they could get in the pasture, which we were told was rather poor, the farm being one that has been mismanaged and run down. Yet their weight is 1,200 pounds each.

"'Dowager' is seven years old, and began her flow of milk on the 19th of June. The quantity given each month was,—

First month, ...	1,434 pounds.
Second "	1,565 "
Third "	1,403 "
Three months' yield, ...	4,402 pounds.

"'Crown Princess' is six years old, and began her flow of milk on the 27th of April, with the following result:

First month, ...	1,707½ pounds.
Second "	2,063 "
Third "	1,697½ "
Fourth "	1,471 "
Four months' yield, ...	6,939 pounds.

"'Fraulein'" is five years old, and began her flow of milk on the 10th of May. The monthly yield was,—

First month, ...	1,873½ pounds.
Second "	1,558 "
Third "	1,021 "
Fourth "	1,009 "
Four months' yield, ...	5,461½ pounds.

"If we call three months, 91 days, and four months, 122 days, we find the following average daily yield:—

'Dowager,' daily yield, ...	48.37 pounds.
'Crown Princess,' daily yield, ...	56.87 "
'Fraulein,' daily yield, ...	44.76 "

"This is at least double the yield of ordinary dairy cows.

"From the facts before us, it would seem that the Holstein cows must prove very valuable and popular in the dairy. Whether they take on flesh readily, and are therefore profitable for beef (as is the case with the Shorthorns), when they cease to be useful in the dairy, we are not informed."

The certificate accompanying the importation describes the animals as pure-bred, and from renowned herds in Friesland.

Another importation has been made by Gen. Wm. S. Tilton, Deputy Governor of the National Military Asylum, at Togus, near Augusta, Me., which consisted of a bull, cow and heifer. The writer has not been able to make an examination of the animals of this importation in person, but an inspection of the documents forwarded by the consignor, proves them to have been carefully selected from some of the best herds in East Friesland, and from pure-bred stock. Gen. Tilton writes that the animals were purchased under instructions to procure the finest specimens of the breed, with no regard to price, and he describes them as follows:—

Bull "Ploen," black-and-white, 2 years old; weight, 1,022 lbs.; girth, 5 feet 10 inches; height, 49½ inches.

Cow "Itzehoe," black-and-white, 5 years old; weight, 1,092 lbs.; girth, 6 feet 10 inches; height, 52½ inches.

Heifer "Altona," black-and-white, 3 years old; weight, 1,174 lbs.; girth, 6 feet 10 inches; height, 48½ inches.

His correspondent in Frankfort, writing at the time of the shipment of the cattle says: "I think that the cattle must be very satisfactory, as they appear to be very fine, and were very much admired by every one who saw them. A few days since I visited the farm of Baron Rothschild, who prides himself upon his stock of cows. I found that he had principally Alderney and Holland stock. The steward informed me that the Holstein were the same as the Holland, except that the former generally had more flesh. I think you will regard this point of difference in your favor, as you desired cattle for beef as well as for milk.

"There are in the vicinity of Frankfort several other very fine and large stables of cows, in which however the Alderney predominate. The owners all speak highly of the Holstein stock, however, and have many cows of that breed."

The animals were brought to this country in the Steamship Saxonia, and landed in New York in September, 1871. Their fine appearance and quality exceeded my anticipations.

The cow "Itzehoe" dropped a bull-calf on the 26th of October, and the heifer "Altona" dropped twins on the 7th of December. They gave a large quantity of milk, an account of which was kept for six or eight weeks, with a constant average product of thirty-five pounds a day to each cow. The greatest day's yield was forty-three and one-half pounds. As I wish to keep these cows in a state of good health for breeding, I have kept them upon ordinary feed.

It is understood that two other importations of cattle have recently been received in this country from North Holland, some of which would doubtless have been entitled to notice and registry in the Holstein Herd Book, provided proper application had been made by their importers or owners. Having no description or statements relative to these two invoices of cattle, we can only allude to the matter—remarking that possibly our next volume may contain a full account of one or both importations.

Taking, as a basis, some statistical facts and calculations in "Allen's American Cattle," and "Willard's Practical Dairy Husbandry," it is estimated that the value of the neat stock in the country at the present time is not far from $1,200,000,000. Included in this valuation are about 11,200,000 cows, yielding annually 43,680,000,000 pounds of milk; the aggregate value of which, in butter and cheese, together with that portion consumed as food, cannot fall short of $400,000,000.

In view of this enormous interest, the importance of accurate knowledge in regard to the relative qualities of different breeds of dairy stock is obvious. The foregoing estimate is based upon an average yield of 1,800 quarts of milk per annum from each cow. If the average production could be raised to 2,000 quarts per cow, which would be comparatively but a slight gain, the country would be enriched thereby more than $40,000,000 annually. Now, if it can be demonstrated that any given breed produce an amount of milk largely in excess of the average yield, and also in excess of any other breed, we have at once before us a visible means of improvement; and in order, therefore, to show the superiority of the Holstein race in the production of milk, reference is had to some comparative experiments by scientific men in Europe, gleaned from a paper upon dairy husbandry, published by John H. Klippart, Secretary Ohio State Board of Agriculture, who visited Germany, Holland and Holstein in 1865.

Premising that the terms "Holland Cow" or "Hollanders," is understood to be synonymous with the terms "Holstein Cow" or Holsteins, as recognized in this country, at this time, we give the results of some experiments illustrating our point.

Professor Lehmann, in charge of an experimental Agricultural Station, at Pommritz, selected cows from both the Shorthorn and Holland races that had passed their sixth year, but were not so old that their age affected the flow of milk. The experiment of milking commenced July 31st, 1866, and terminated July 30th, 1867, being precisely 365 days: during which time there was yielded by four Shorthorns, 27,204 pounds of milk, or an average of 6,801 pounds per cow; while four Hollanders yielded 32,136 pounds, or an average of 8,034 pounds per cow. Highest yield by Shorthorn, 7,643 pounds; highest yield by Hollanders, 9,411 pounds. The same food, keeping, care and surroundings, which in the Shorthorns produced 100 pounds of milk, produced 118$\frac{4}{10}$ in the Hollanders, or fully 18 per cent. in favor of the latter breeds as milk producers.

At the Agricultural Academy, at Eldena, many experiments were made in feeding and milking cows, &c., and very precise accounts were kept of the product of every cow, as well as the expense of keeping her; and it was found that three Ayrshire cows averaged 2,247 quarts of milk per cow, while twenty-two Holland cows averaged 4,437 quarts per cow: highest yield by Ayrshire, 2,811 quarts; highest yield by Hollander, 5,677 quarts. The Hollanders consume about five pounds of the equivalent of hay for every quart of milk yielded; the Ayrshires, nine pounds of hay for one quart of milk. Another series of experiments, conducted by Villeroy, resulted in showing that 100 pounds of hay produced, in Hollanders 28.92 quarts of milk; in Devons, 19.13 quarts; and in Herefords, 15.97 quarts.

"Baron Ockel, in Frankenfelde, made a comparative experiment with Ayrshires and Hollanders; the average weight of the Ayrshires being 806 pounds, and that of the Hollanders 1,016 pounds. The experiment showed that the Ayrshires consumed $3\frac{3}{10}$ pounds of hay for every 100 pounds of live weight, while the Hollanders consumed $2\frac{8}{10}$ only. Of the amount of food consumed, one-sixtieth of their live weight only was required to keep the Hollanders in their normal state, while it required one-fiftieth of their live weight, in food of hay equivalents, to keep the Ayrshires in their normal condition." Baron Ockel also made another experiment with four Holland cows; the two heaviest of which weighed 2,112 pounds, and the lighter ones 1,537 pounds. He placed them in two groups,—the heavy ones forming one group and the light ones forming the other,—and continued the experiment during sixteen days, the food being correctly weighed to each when fed; the result showing "that heavy cows of the same breed consume relatively less food than the lighter ones, and at the same time yield a greater return for it." Mention is made of one dairy of 190 head of Hollanders, which averaged 4,076 quarts of milk a cow per annum.

"The German dairymen never purchase a cow in milk, for the dairy. They purchase either a heifer in calf, or a cow in calf, but never purchase a cow with a calf." Mr. Ree, an extensive dairy farmer, in Holstein, stated "that every *cow* offered for sale, was a *bad* cow; no man would sell a good cow."

There having been some criticisms in relation to the name or designation of the race of cattle under consideration, it is proper to state that, in view of the fact that some confusion of terms had existed on this point,—the cattle having been variously called "Holstein," "Dutch," "Holstein or Dutch," "Dutch or Holstein," "Holland," "North German," &c.,—it was deemed advisable, by breeders of the stock, to fix upon some definite name; and inasmuch as the best authorities describe the progenitors of the cattle in question as having originated in the Duchy of Holstein, and thence been distributed through the adjacent provinces, "*Holstein*" was considered to be the most proper designation, and the Holstein Breeders' Association resolved to avoid further confusion of terms by adhereing to that as eminently fit and appropriate.

The position of the Holstein breeders, in regard to the name of *Holstein*, is understood to be analogous to that of the Devon breeders, and Jersey breeders, in regard to the respective names, *Devon* and *Jersey*. It is not claimed that the best specimens of the Holstein race are now in the Duchy of Holstein, neither is it probable that the best Jerseys are to be had in the Island of Jersey, or the best Devons in Devonshire. The question, therefore, whether the best of the race are, or are not, found in Holstein at the present time, has no significance.

The early English writers upon the subject of breeds of cattle, state, as before quoted, that "the finest cattle of the North of Europe have been derived from Holstein," and thence, especially "the finest of the Dutch breeds had themselves been derived;" and the German writer, also, before referred to, says, "although the race has been most fully developed and attained to the greatest consequence in North Holland, *the original stock was by no means bred in Holland, but in Holstein*, whence it spread itself over the North of Germany and Holland."

Klippart says: "I found *large black-and-white cattle* in Holstein recognized as 'Ditmarsh' cattle; then I found another *large black-and-white cattle* which I could barely tell from Shorthorns by their size and form, and which were variously called 'Holland' or 'Oldenbergers'; then there is another *black-and-white breed* known as the 'Breitenbergers'; and in Holland there are two strains of *black-and-white* cattle, famous as milkers, and known as 'Beemsters,' and the other famous milkers known as 'Frieslanders.' Now all these breeds differ from each other—*they are all black-and-white, all large, and all good milkers;*" and he says, "in the cattle markets in Smithfield (London), Hamburg, Berlin, &c., you will find all these breeds," and others, "called and recognized in the aggregate as 'Holsteiners'—that is, cattle from Holstein."

Dr. Detmers, a native of Oldenburg, speaks of the "Lowland breed of cattle in Holland, Friesland, and the northern part of the Grand Duchy of Oldenburg," and "the excellent qualities and the extraordinary size of that magnificent breed"; and Dr. Bocking, another native German, says, the best representatives of the breed "being found only on the borders of North Holland, but very extensively in Ost Friesland and the Diet Marshes, the name Dutch cattle is hardly the most appropriate. In their country they are called Marsh cattle; and nine-tenths of the whole being raised in Germany, in a mercantile point of view, and pointing to the centre of their climate and soil, would it not be better to call them North Germans?" . . . . They are, he says, "the breed of my native country . . . . and the experience of my whole life makes me believe that, for our dairy purposes, no blood can equal them." Another gentleman, a native of Mecklenburg, in a letter now before me, says: "I am very glad to observe that more correct views regarding the name of the Holstein cattle are going to prevail. In my native country that breed is principally esteemed, and is imported from Holstein. We call every cow a Hollander cow, in case it belongs to the large herds of the great estate owners. These gentlemen often keep from 200 to 800 head of cows, and rent them (or the produce of them) to a person that is called a 'Hollander'—meaning natives of Holland, who came in former times to hire as herdsmen in dairies. This name remained attached to those in charge of large herds of cattle for dairy purposes; the name 'Hollanderie,' to a large dairy: *hence originated the name of 'Holland cow'*—meaning *dairy* cow."

Holstein has large bottom lands with rich pastures, upon which are raised the *famous hardy, black-and-white cows especially adapted for dairy purposes.* Every spring, thousands of the Holstein heifers are driven to the fields of Northern Germany and Holland, where people find it more profitable to buy heifers than to raise them; and the name of the breed got confused, so that the name "Holland cow" was here translated into "Dutch cow." . . I think this breed will be more adapted for the climate of America than the English breeds, the latter being not so hardy as the former.

Flint, in his "Agriculture of Massachusetts, for 1863," in reference to these cattle, says: "the same kind of soil, in fact, the same general features, characterize the whole coast-line from Flanders and Belgium, along round the shores of Holland and Hanover, as far as the Elbe, on which lies the great city of Hamburg, and so still farther to the north and east, taking in Holstein and Schleswig. It is a magnificent stretch of marsh land. If we look over the races of cattle that have grown up as natives of these low marshes, or netherlands, different as they are in many respects, we shall find *the same general characteristics running through them all.* Perhaps the Dutch may be taken as the most prominent type of these lowland or marsh races. It is found in its greatest purity in North Holland, Friesland and Groningen; but is much more widely spread than we should conclude from the size of these provinces on the map."

Now, it appears by the accounts of these various writers that there is, existing in the various provinces named, a race of cattle which, owing probably to different care and treatment, now present some dissimilarity in different localities, but which, in all the various places, possess *the same general characteristics,*—all being of *large size, black-and-white, and all extraordinary milkers;* and although the local names by which the cattle are known in the several localities where they are raised, are doubtless proper and appropriate, yet the breeders of the stock in this country need a distinctive name applicable to the general stock, whether from North Holland, Friesland, Oldenburg, or Holstein; and it is submitted that, for the reasons before given, the term *Holstein,* as recognized and adopted by the Holstein Breeders'

Association, is the most fit and proper designation; and it is gratifying to know that the prominent agricultural societies and agricultural journals in the country acquiesce in the action of the Association, and use the term Holstein in their publications.

All accounts concur in admitting that the best specimens of the race exist in North Holland, Friesland and Oldenburg: hence the wisdom of the resolution defining the limits embraced in those provinces as the territory from which importations of animals would be recognized as Holsteins by the Association; and if the Holstein cattle are, as we claim, a superior race, worthy of importation and propagation in this country, then the importance and value of a Holstein Herd Book is obvious; and the Holstein Herd Book, it may be said, stands upon the same footing in this country as the Jersey and Ayrshire Herd Books. It is intended to contain records of all imported animals, and pedigrees of all animals tracing their lineage to imported stock, as defined in the by-laws and resolves of "The Association of Breeders of Thoroughbred Holstein Cattle."

In concluding this sketch, the writer deems it proper to say that, while confessing a very decided partiality for the Holstein race, he disclaims having any intention of attempting to detract from the merits of any other breed of cattle. In this broad country there is room for all, and every breed may find an appropriate place. The majestic Shorthorn, commanding at all times, and in all places, unqualified admiration, finds, in the blue-grass pastures of the Western States, and in other localities where feed is superabundant, his meat-producing capacity appreciated and unquestioned. And we are not unmindful of the exquisite beauty of the fawn-like Jersey in her appropriate place upon the lawn of the affluent or aristocratic country gentleman, or disposed to undervalue her productions of rich cream and butter. The compact little Devon, the Hereford, and the Ayrshire, are all, doubtless, valuable in suitable locations and under favorable circumstances; yet, admitting the value of all these, and other meritorious breeds, it is claimed that the Holsteins are superior in their *combination of desirable properties*,—producing milch cows that yield more than an average quantity of butter, and incomparably more milk, especially adapted to cheese-making as well as for family use, than any other breed; working oxen, large, strong and symmetrical, and at the same time hardy, spirited and tractable; and withal, being exceedingly docile, quiet, and good feeders, with great aptitude to take on flesh, well adapted for the production of beef.

The magnificent appearance presented by a herd of Holsteins, when grouped in pasture or upon the lawn, is not the least of their valuable characteristics,—excelling the Shorthorns in size, and approximating them in form, the snow white and jet black colors, "mottled in picturesque inequalities of proportion over their bodies," present a striking contrast, and a conspicuous and beautiful feature in the landscape.

HOLSTEIN HERD RECORD.

BULLS.

No. 1. Amsterdam.

Black and white; calved April 12, 1867; bred by CHARLES HOUGHTON, Putney, Vt.; the property of GEORGE G. LOBDELL, Wilmington, Del. Sire, Van Tromp (50); imported from North Holland by WINTHROP W. CHENERY, 1861. Dam, Midwould 2d (25); by 2d Dutchman (37); grandam, Lady Midwould (17); imported from North Holland by WINTHROP W. CHENERY, 1861.

No. 2. Bismarck.

Black, with a little white; calved March 15, 1870; bred by CHARLES HOUGHTON, Putney, Vt.; the property of C. C. WALWORTH, Monticello, Iowa. Sire, Van Tromp (50); imported from North Holland by WINTHROP W. CHENERY, 1861. Dam, Midwould 6th (29); by Zuider Zee 2d (57); grandam, Lady Midwould (17); imported from North Holland by WINTHROP W. CHENERY, 1861.

No. 3. Bleecker.

White and black; calved March 29, 1872; bred by and the property of GERRIT S. MILLER, Peterboro', Madison County, N. Y. Sire, Hollander (20); imported from West Friesland by GERRIT S. MILLER, 1869. Dam, Fraulein (9); imported from West Freisland by GERRIT S. MILLER, 1869.

No. 4. Denmark.

Black and white; calved October 26, 1871; imported in cow Itzehoe (13); from East Friesland by, and the property of, NATIONAL MILITARY ASYLUM, Wm. S. Tilton, Dep. Gov., Augusta, Me.

No. 5. Duke of Belmont.

Black and white; calved February 22, 1868; bred by and the property of WINTHROP W. CHENERY, Belmont, Mass. Sire, Van Tromp (50). Dam, Maid of Opperdoes (22). Both imported from North Holland by WINTHROP W. CHENERY, 1861.

No. 6. Duke of Holstein.

Black and white; calved August 26, 1863; bred by and the property of WINTHROP W. CHENERY, Belmont, Mass. Sire, Hollander (19). Dam, Lady Midwould (17). Both imported from North Holland by WINTHROP W. CHENERY, 1861.

No. 7. Dutchman.

Black and white; calved in 1855; bred in North Holland, thence imported by WINTHROP W. CHENERY, Belmont, Mass., 1857. Slaughtered in 1861, leaving no pure-bred progeny excepting his son, 2d Dutchman (37).

No. 8. Earl of Middlesex.

White, with black spots; calved March 12, 1868; bred by WINTHROP W. CHENERY, Belmont, Mass.; the property of DOYLESTOWN AGRICULTURAL AND MECHANICS' INSTITUTE, Doylestown, Bucks County, Penn. Sire, Van Tromp (50). Dam, Zuider Zee (62); both imported from North Holland by WINTHROP W. CHENERY, 1861.

No. 9. Fifth Duke of Belmont.

Black, with white marks; calved March 31, 1872; bred by and the property of WINTHROP W. CHENERY, Belmont, Mass. Sire, Highland Chief (18); grandsire, Van Tromp (50); imported from North Holland by WINTHROP W. CHENERY, 1861. Dam, Opperdoes 3d (39); by Hollander (19); grandam, Maid of Opperdoes (22); imported from North Holland by WINTHROP W. CHENERY, 1861.

No. 10. Fifth Highland Chief.

Black, with white marks; calved April 10, 1872; bred by and the property of WINTHROP W. CHENERY, Belmont, Mass. Sire, Duke of Belmont (5); grandsire, Van Tromp (50); imported from North Holland by WINTHROP

W. CHENERY, 1861. Dam, Midwould 2d (25); by 2d Dutchman (37); grandam, Lady Midwould (17); imported from North Holland by WINTHROP W. CHENERY, 1861.

No. 11. Fifth Lord of Texelaar.

Black, with white marks; calved September 10, 1871; bred by and the property of WINTHROP W. CHENERY, Belmont, Mass. Sire, Duke of Holstein (6); grandsire, Hollander (19); imported from North Holland by WINTHROP W. CHENERY, 1861. Dam, Texelaar 3d (52); by 2d Dutchman (37); grandam, Texelaar (51); imported from North Holland by WINTHROP W. CHENERY, 1861.

No. 12. Fourth Duke of Belmont.

White, with black marks; calved March 23, 1872; bred by and the property of WINTHROP W. CHENERY, Belmont, Mass. Sire, Highland Chief (18); grandsire, Van Tromp (50); imported from North Holland by WINTHROP W. CHENERY, 1861. Dam, Opperdoes 2d (38), by 2d Dutchman (37); grandam, Maid of Opperdoes (22); imported from North Holland by WINTHROP W. CHENERY, 1861.

No. 13. Fourth Earl of Middlesex.

White, with black spots; calved March 4, 1871; bred by WINTHROP W. CHENERY, Belmont, Mass.; the property of H. C. HOFFMAN, Horse Heads, Chemung County, N. Y. Sire, Texelaar 6th (44); grandsire, Zuider Zee 2d (57); imported from North Holland by WINTHROP W. CHENERY, 1861. Dam, Zuider Zee (62); also imported from North Holland by WINTHROP W. CHENERY, 1861.

No. 14. Fourth Highland Chief.

Black and white; calved May 15, 1871; bred by WINTHROP W. CHENERY, Belmont, Mass.; the property of the MASSACHUSETTS AGRICULTURAL COLLEGE, Amherst, Mass. Sire, Texelaar 6th; (44) grandsire, Zuider Zee 2d (57); imported from North Holland by WINTHROP W. CHENERY, 1861. Dam, Midwould 9th (32); by Van Tromp (50); grandam, Lady Midwould (17); imported from North Holland by WINTHROP W. CHENERY, 1861.

No. 15. Fourth Lord of Texelaar.

White, with black spots; calved June 9, 1871; bred by WINTHROP W. CHENERY, Belmont, Mass.; the property of B. E. STEWART, North Yamhill,

Yamhill County, Oregon. Sire, Highland Chief (18); grandsire, Van Tromp (50); imported from North Holland by WINTHROP W. CHENERY, 1861. Dam, Texelaar 11th (58); by Van Tromp (50); grandam, Texelaar 3d (52); by 2d Dutchman (37); g grandam, Texelaar (51); imported from North Holland by WINTHROP W. CHENERY, 1861.

No. 16. Goldfinder.

Black and white; calved November 20, 1870; bred by and the property of S. B. EMERSON, Mountain View, Santa Clara County, Cal. Sire, Opperdoes 4th (29); grandsire, Zuider Zee 2d (57); imported from North Holland by WINTHROP W. CHENERY, 1861. Dam, Opperdoes 8th (40); by Van Tromp (50); grandam, Maid of Opperdoes (22); imported from North Holland by WINTHROP W. CHENERY, 1861.

No. 17. Hamilcar.

Black and white; calved March 30, 1870; imported in cow Crown Princess (6), from West Freisland, by and the property of GERRIT S. MILLER, Peterboro', Madison County, N. Y., 1869.

No. 18. Highland Chief.

White and black; calved April 29, 1869; bred by WINTHROP W. CHENERY, Belmont, Mass.; the property of JOHN H. COMER, Goshen, N. Y. Sire, Van Tromp (50); imported from North Holland by WINTHROP W. CHENERY, 1861. Dam, Midwould 6th (30); by Zuider Zee 2d (57); grandam, Lady Midwould (17); imported from North Holland by WINTHROP W. CHENERY, 1861.

No. 19. Hollander (Chenery's).

Black and white; calved in 1860; bred in North Holland; thence imported by WINTHROP W. CHENERY, Belmont, Mass., 1861; now the property of SAMUEL FAIRBANK, Oakham, Mass.

No. 20. Hollander (Miller's).

Black and white; calved in 1867; bred in West Freisland; thence imported by and the property of GERRIT S. MILLER, Peterboro', N. Y., 1869.

No. 21. Horace Greeley.

Black, with a little white; calved December 7, 1871; imported in cow Altona (2), from East Freisland, by NATIONAL MILITARY ASYLUM, Wm. S. Tilton, Dep. Gov., Augusta, Me., 1871; the property of WINTHROP W. CHENERY, Belmont, Mass.

No. 22. Kaiser.

Black and white; calved March 11, 1871; bred by and the property of CHARLES HOUGHTON, Putney, Vt. Sire, Amsterdam (1); grandsire, Van Tromp (50); imported from North Holland by WINTHROP W. CHENERY, 1861. Dam, Midwould 6th (30); by Zuider Zee 2d (57); grandam, Lady Midwould (17); imported from North Holland by WINTHROP W. CHENERY, 1861.

No. 23. Lord of Texelaar.

Black, with white spots; calved January 18, 1870; bred by WINTHROP W. CHENERY, Belmont, Mass.; the property of ONEIDA COMMUNITY, Oneida, N. Y. Sire, 3d Dutchman (46); grandsire, 2d Dutchman (37); g. grandsire, Dutchman (7); imported from North Holland by WINTHROP W. CHENERY, 1857. Dam, Texelaar (51); also imported from North Holland by WINTHROP W. CHENERY, 1861.

No. 24. Merrimac.

White and black; calved April 20, 1871; bred by and the property of W. A. RUSSELL, Lawrence, Mass. Sire, Zuider Zee 2d (57); imported from North Holland by WINTHROP W. CHENERY, 1861. Dam, Midwould 4th (26); by Hollander (19); grandam, Lady Midwould (17); imported from North Holland by WINTHROP W. CHENERY, 1861.

No. 25. Midwould 4th.

White, with black spots; calved July, 1867; bred by THOMAS S. LANG, North Vassalboro, Me.; the property of C. C. WALWORTH, Monticello, Iowa. Sire, Duke of Holstein (6); grandsire, Hollander (19); imported from North Holland by WINTHROP W. CHENERY, 1861. Dam, Zuider Zee 3d (63); by Hollander (19); grandam, Zuider Zee (62); imported from North Holland by WINTHROP W. CHENERY, 1861.

No. 26. Midwould 5th.

White, with black spots; calved July, 1868; bred by JAMES S. MUNROE, Lexington, Mass.; the property of C. C. WALWORTH, Monticello, Iowa. Sire, Duke of Holstein (6); grandsire, Hollander (19) imported from North Holland by WINTHROP W. CHENERY, Belmont, Mass., 1861. Dam, Zuider Zee 3d (63); by Hollander (19); grandam, Zuider Zee (62); imported from North Holland by WINTHROP W. CHENERY, 1861.

No. 27. Midwould 7th.

Black and white; calved May 5, 1866; bred by and the property of WINTHROP W. CHENERY, Belmont, Mass. Sire, Van Tromp (50); imported from North Holland by WINTHROP W. CHENERY, 1861. Dam, Midwould 2d (25); by 2d Dutchman (37); grandam, Lady Midwould (17); imported from North Holland by WINTHROP W. CHENERY, 1861.

No. 28. Monticello Duke.

Black and white; calved December 25, 1871; bred by and the property of C. C. WALWORTH, Monticello, Iowa. Sire, 2d Duke of Belmont (36); grandsire, Van Tromp (50); imported from North Holland by WINTHROP W. CHENERY, 1861. Dam, Duchess of Holstein (8); by Duke of Holstein (6); grandam, Zuider Zee 3d (63); by Hollander (19); g. grandam Zuider Zee (62); imported from North Holland by WINTHROP W. CHENERY, 1861.

No. 29. Opperdoes 4th.

Black and white; calved June 15, 1865; bred by WINTHROP W. CHENERY, Belmont, Mass.; the property of S. B. Emerson, Mountain View, Santa Clara County, Cal. Sire, Zuider Zee 2d (57); imported from North Holland by WINTHROP W. CHENERY, 1861. Dam, Opperdoes 2d (38); by 2d Dutchman (37); grandam, Maid of Opperdoes (22); imported from North Holland by WINTHROP W. CHENERY, 1861.

No. 30. Opperdoes 5th.

White, with black ears; calved July 23, 1865; bred by WINTHROP W. CHENERY, Belmont, Mass.; the property of WM. SMITH, Proctorsville, Vt. Sire, Hollander (19). Dam, Maid of Opperdoes (22); both imported by WINTHROP W. CHENERY, 1861.

No. 31. Opperdoes 6th.

Black, with white marks; calved June 7, 1866; bred by and died the property of WINTHROP W. CHENERY, Belmont, Mass. Sire, Hollander (19); dam, Maid of Opperdoes (22); both imported from North Holland by WINTHROP W. CHENERY, 1861.

No. 32. Opperdoes 7th.

Black and white; calved July 23, 1866; bred by WINTHROP W. CHENERY, Belmont, Mass.; the property of HENRY WATERMAN, North Kingston, R. I. Sire, Van Tromp (50); imported from North Holland by WINTHROP W. CHENERY. Dam, Opperdoes 2d (38); by 2d Dutchman (37); grandam, Maid of Opperdoes (22); imported from North Holland by WINTHROP W. CHENERY, 1861.

No. 33. Opperdoes 12th.

Black and white; calved November 11, 1868; bred by WINTHROP W. CHENERY, Belmont, Mass.; the property of S. B. EMERSON, Mountain View, Santa Clara County, California. Sire, Midwould 7th (27); grandsire, Van Tromp (50); imported from North Holland by WINTHROP W. CHENERY, 1861. Dam, Opperdoes 2d (38); by 2d Dutchman (37); grandam, Maid of Opperdoes (22); imported from North Holland by WINTHROP W. CHENERY, 1861.

No. 34. Ploen.

Black and white; calved in 1870; bred in East Friesland; thence imported by and the property of NATIONAL MILITARY ASYLUM, Wm. S. Tilton, Dep. Gov., Augusta, Me., 1871.

No. 35. Rip Van Winkle.

Black and white; calved April 18, 1870; imported in cow Fraulein (9), from West Friesland by and the property of GERRIT S. MILLER, Peterboro', N. Y., 1869.

No. 36. Second Duke of Belmont.

Black and white; calved December 11, 1868; bred by WINTHROP W. CHENERY, Belmont, Mass.; the property of THOMAS J. DUNBAR, Boston,

Mass. Sire, Van Tromp (50); imported from North Holland by WINTHROP W. CHENERY, 1861. Dam, Opperdoes 3d (39); by Hollander (19); grandam, Maid of Opperdoes (22); imported from North Holland by WINTHROP W. CHENERY, 1861.

No. 37. Second Dutchman.

Black and white; calved February 15, 1859; bred by WINTHROP W. CHENERY, Belmont, Mass.; the property of CHARLES W. CUSHING, Hingham, Mass. Sire, Dutchman (7); dam, Lady Rutten (19); both imported from North Holland by WINTHROP W. CHENERY, 1857.

Second Dutchman is the only pure-blood animal now living descended from the "Chenery importations" of 1852, 1857, and 1859.

No. 38. Second Earl of Middlesex.

Black and white; calved February 10, 1869; bred by and the property of WINTHROP W. CHENERY, Belmont, Mass. Sire, Van Tromp (50); dam, Zuider Zee (62); both imported from North Holland by WINTHROP W. CHENERY, 1861.

No. 39. Second Highland Chief.

Black and white; calved June 1, 1869; bred by and the property of WINTHROP W. CHENERY, Belmont, Mass. Sire, Van Tromp (50); dam, Lady Midwould (17); both imported from North Holland by WINTHROP W. CHENERY, 1861.

No. 40. Second Lord of Texelaar.

White and black; calved July 13, 1870; bred by WINTHROP W. CHENERY, Belmont, Mass.; the property of WILLIAM STECKEL, Doylestown, Penn. Sire, 3d Dutchman (46); grandsire, 2d Dutchman (37); g. grandsire, Dutchman (7); imported from North Holland by WINTHROP W. CHENERY, 1857. Dam, Texelaar 10th (57); by Opperdoes 4th (29); grandam, Texelaar (51); imported from North Holland by WINTHROP W. CHENERY, 1861.

No. 41. Sixth Highland Chief.

White, with black spots; calved April 9, 1872; bred by and the property of WINTHROP W. CHENERY, Belmont, Mass. Sire, Duke of Belmont (5);

grandsire, Van Tromp (50); imported from North Holland by WINTHROP W. CHENERY, 1861. Dam, Midwould 9th (32); by Van Tromp (50); grandam, Lady Midwould (17); imported from North Holland by WINTHROP W. CHENERY, 1861.

No. 42. Sixth Lord of Texelaar.

Black and white; calved April 5, 1872; bred by and the property of WINTHROP W. CHENERY, Belmont, Mass. Sire, Duke of Holstein (6); grandsire, Hollander (19); imported from North Holland by WINTHROP W. CHENERY, 1861. Dam, Texelaar (51); imported from North Holland by WINTHROP W. CHENERY, 1861.

No. 43. Texelaar 4th.

Black and white; calved March 4, 1864; bred by WINTHROP W. CHENERY, Belmont, Mass.; died the property of JOHN M. QUIGG, Spencer, Tioga County, N. Y. Sire, Hollander (19); dam, Texelaar (51); both imported from North Holland by WINTHROP W. CHENERY, 1861.

No. 44. Texelaar 6th.

Black and white; calved December 22, 1865; bred by and died the property of WINTHROP W. CHENERY, Belmont, Mass. Sire, Zuider Zee 2d (57); imported from North Holland by WINTHROP W. CHENERY, 1861. Dam, Texelaar 3d (52), by 2d Dutchman (37); grandam, Texelaar (51); imported from North Holland by WINTHROP W. CHENERY, 1861.

No. 45. Third Duke of Belmont.

Black and white; calved April 26, 1870; bred by WINTHROP W. CHENERY, Belmont, Mass.; the property of LUCIAN D. TROW, Hardwick, Mass. Sire, Earl of Middlesex (8); grandsire, Van Tromp (50); imported from North Holland by WINTHROP W. CHENERY, 1861. Dam, Opperdoes 3d (39), by Hollander (19); grandam, Maid of Opperdoes (22); imported from North Holland by WINTHROP W. CHENERY, 1861.

No. 46. Third Dutchman.

Black and white; calved August 5, 1867; bred by WINTHROP W. CHENERY, Belmont, Mass.; the property of the ONEIDA COMMUNITY, Oneida,

N. Y. Sire, 2d Dutchman (37); grandsire, Dutchman (7); imported from North Holland by WINTHROP W. CHENERY, 1857. Dam, Opperdoes 3d (39); by Hollander (19); grandam, Maid of Opperdoes (22); imported from North Holland by WINTHROP W. CHENERY, 1861.

No. 47. Third Earl of Middlesex.

Black and white; calved February 10, 1869; bred by and the property of WINTHROP W. CHENERY, Belmont, Mass. Sire, Van Tromp (50); dam, Zuider Zee (62); both imported from North Holland by WINTHROP W. CHENERY, 1861.

No. 48. Third Highland Chief.

Black and white; calved July 1, 1870; bred by WINTHROP W. CHENERY, Belmont, Mass.; the property of JOHN CUMMINGS, Woburn, Mass. Sire, Duke of Belmont (5); grandsire, Van Tromp (50); imported from North Holland by WINTHROP W. CHENERY, 1861. Dam, Midwould 2d (25); by 2d Dutchman (37); grandam, Lady Midwould (17); imported from North Holland by WINTHROP W. CHENERY, 1861.

No. 49. Third Lord of Texelaar.

Black and white; calved August 17, 1870; bred by and the property of WINTHROP W. CHENERY, Belmont, Mass. Sire, Earl of Middlesex (8); grandsire, Van Tromp (50); imported from North Holland by WINTHROP W. CHENERY, 1861. Dam, Texelaar 3d (52); by 2d Dutchman (37); grandam, Texelaar (51); imported from North Holland by WINTHROP W. CHENERY, 1861.

No. 50. Van Tromp.

Black and white; calved March 20, 1862; imported in cow Texelaar (51), from North Holland by and died the property of WINTHROP W. CHENERY, Belmont, Mass., 1861.

No. 51. Van Tromp, Jr.

Black and white; calved April 23, 1870; bred by and the property of WINTHROP W. CHENERY, Belmont, Mass. Sire, Van Tromp (50); dam, Lady Midwould (17); both imported from North Holland by WINTHROP W. CHENERY, 1861.

No. 52. Van Tromp 2d.

Black and white; calved February 4, 1869; bred by WINTHROP W. CHENERY, Belmont, Mass.; the property of THOMAS B. WALES, Jr., South Framingham, Mass. Sire, Van Tromp (50); imported from North Holland by WINTHROP W. CHENERY, 1861. Dam, Texelaar 3d (52); by 2d Dutchman (37); grandam, Texelaar (51); imported from Holland by WINTHROP W. CHENERY, 1861.

No. 53. Van Tromp 3d.

Black and white; calved March 4, 1872; bred by and the property of THOMAS B. WALES, Jr., South Framingham, Mass. Sire, Van Tromp 2d (52); grandsire, Van Tromp (50); imported from North Holland by WINTHROP W. CHENERY, 1861. Dam, Texelaar 12th (59); by 3d Dutchman (46); grandam, Texelaar 8th (55); by Zuider Zee 2d (57); g. g. dam, Texelaar (51); imported from North Holland by WINTHROP W. CHENERY, 1861.

No. 54. Van Tromp 4th.

Black and white; calved May 10, 1872; bred by and the property of THOMAS B. WALES, Jr., South Framingham, Mass. Sire, Van Tromp 2d (52); grandsire, Van Tromp (50); imported from North Holland by WINTHROP W. CHENERY, 1861. Dam, Maid of Opperdoes (22); imported from North Holland by WINTHROP W. CHENERY, 1861.

No. 55. Vermont.

White and black; calved March 3d, 1868; bred by CHARLES HOUGHTON, Putney, Vt.; the property of LEVI CHAMPION, East Canaan, N. H. Sire, Texelaar 6th (44); grandsire, Zuider Zee 2d (57); imported from North Holland by WINTHROP W. CHENERY, Belmont, Mass., 1861. Dam, Midwould 2d (25); by 3d Dutchman (37); grandam, Lady Midwould (17); imported from North Holland by WINTHROP W. CHENERY, 1861.

No. 56. William.

Black, with a little white; calved May 10, 1870; bred by and the property of CHARLES HOUGHTON, Putney, Vt. Sire, 3d Dutchman (46); grandsire, 2d Dutchman (37); g. grandsire, Dutchman (7); imported from North Holland by WINTHROP W. CHENERY, Belmont, Mass., 1857. Dam, Texelaar 8th (55); by Zuider Zee 2d (57); grandam, Texelaar (51); imported from North Holland by WINTHROP W. CHENERY, 1861.

No. 57. Zuider Zee 2d.

White and black; calved March 7, 1862; imported in cow Zuider Zee (62), from North Holland, by WINTHROP W. CHENERY, Belmont, Mass., 1861; the property of W. A. RUSSELL, Lawrence, Mass.

No. 58. Zuider Zee 3d.

Black and white; calved May 16, 1868; bred by and the property of W. A. RUSSELL, Lawrence, Mass. Sire, Zuider Zee 2d (57); imported from North Holland by WINTHROP W. CHENERY, 1861. Dam, Lady Oldenburg (18); imported from Oldenburg by Dr. HATCH.

No. 59. Zuider Zee 4th.

Black and white; calved April 21, 1866; bred by WINTHROP W. CHENERY, Belmont, Mass.; the property of O. F. JONES, Wooster, Ohio. Sire, Van Tromp (50). Dam, Zuider Zee (62); both imported from North Holland by WINTHROP W. CHENERY, 1861.

No. 60. Zuider Zee 5th.

White and black; calved April 6, 1868; bred by and the property of W. A. RUSSELL, Lawrence, Mass. Sire, Zuider Zee 2d (57); imported from North Holland by WINTHROP W. CHENERY, Belmont, Mass, 1861. Dam, Midwould 4th (26); by Hollander (19); grandam, Lady Midwould (17); imported from North Holland by WINTHROP W. CHENERY, 1861.

No. 61. Zuider Zee 6th.

Black and white; calved December 10, 1869; bred by and the property of W. A. RUSSELL, Lawrence, Mass. Sire, Zuider Zee 2d (57); imported from North Holland by WINTHROP W. CHENERY, 1861. Dam, Lady Oldenburg (18); imported from Oldenburg by Dr. HATCH.

COWS.

No. 1. Agoo.

Black and white; calved March 15, 1870; imported in cow Dowager (7), from West Friesland, by and the property of Gerrit S. Miller, Peterboro', N. Y., 1869.

No. 2. Altona.

Black and white; calved in 1869; bred in East Friesland; thence imported by and the property of National Military Asylum, Wm. S. Tilton, Dep. Gov. Augusta, Me., 1871.

No. 3. Belle of Essex.

Black and white; calved November 23, 1870; bred by and the property of W. A. Russell, Lawrence, Mass. Sire, Zuider Zee 2d (57); imported from North Holland by Winthrop W. Chenery, Belmont, Mass., 1861. Dam, Midwould 5th (28); by Zuider Zee 2d (57); grandam, Midwould 4th (26); by Hollander (19); g. grandam, Lady Midwould (17); imported from North Holland by Winthrop W. Chenery, 1861.

No. 4. Belle of Spring Ridge.

Black and white; calved April 11, 1871; bred by and the property of William Steckel, Doylestown, Bucks County, Penn. Sire, Earl of Middlesex (8); grandsire, Van Tromp (50); imported from North Holland by Winthrop W. Chenery, 1861. Dam, Texelaar 9th (56); by Van Tromp (50); grandam, Texelaar 3d (52); by 2d Dutchman (37); g. grandam, Texelaar (51); imported from North Holland by Winthrop W. Chenery, 1861.

property
; imported
n, Midwould

No. 5. Bessie.

Black and white; calved May 10, 1866; bred by and the p 17); imported
A. Russell, Lawrence, Mass. Sire, Zuider Zee 2d (57);

North Holland by WINTHROP W. CHENERY, 1861. Dam, Lady Oldenburg (18); imported from Oldenburg by Dr. HATCH.

No. 6. Crown Princess.

Black and white; calved in 1865; bred in West Friesland; thence imported by and the property of GERRIT S. MILLER, Peterboro', N. Y., 1869.

No. 7. Dowager.

Black and white; calved in 1864; bred in West Friesland; thence imported by and the property of GERRIT S. MILLER, Peterboro', N. Y., 1869.

No. 8. Duchess of Holstein.

White and black; calved August 26, 1869; bred by JAMES S. MUNROE, Lexington, Mass.; the property of C. C. WALWORTH, Monticello, Iowa. Sire, Duke of Holstein (6); grandsire, Hollander (19); imported from North Holland by WINTHROP W. CHENERY, 1861. Dam, Zuider Zee 3d (63); by Hollander (19); grandam, Zuider Zee (62); imported from North Holland by WINTHROP W. CHENERY, 1867.

No. 9. Fraulein.

Black and white; calved in 1866; bred in West Friesland; thence imported by and the property of GERRIT S. MILLER, Peterboro', N. Y., 1869.

No. 10. Grand Duchess.

Black and white; calved February 8, 1869; bred by and the property of CHARLES HOUGHTON, Putney, Vt.; sire, Texelaar 6th (44); grandsire, Zuider Zee 2d (57); imported from North Holland by WINTHROP W. CHENERY, 1861. Dam, Midwould 2d (25); by 2d Dutchman (37); grandam, Lady Midwould (17); imported from North Holland by WINTHROP W. CHENERY, 1861.

No. 11. Gretchen.

lack spots; calved September 27, 1870; bred by JAMES S. ton, Mass.; the property of CHARLES HOUGHTON, Putney,

Vt. Sire, Duke of Holstein (6); grandsire, Hollander (19); imported from North Holland by WINTHROP W. CHENERY, 1861. Dam, Zuider Zee 3d (63); by Hollander (19); grandam, Zuider Zee (62); imported from North Holland by WINTHROP W. CHENERY, 1861.

No. 12. Hebe.

Black and white; calved May 7, 1871; bred by and the property of GERRIT S. MILLER, Peterboro', N. Y. Sire, Hollander (20); dam, Fraulein (9); both imported from West Friesland by GERRIT S. MILLER, 1869.

No. 13. Itzehoe.

Black and white; calved in 1867; bred in East Friesland; thence imported by and the property of NATIONAL MILITARY ASYLUM, Wm. S. Tilton, Dep. Gov., Augusta, Me., 1871.

No. 14. Juno.

Black and white; calved April 21, 1871; bred by and died the property of GERRIT S. MILLER, Peterboro', Madison County, N. Y. Sire, Hollander (20); dam, Crown Princess (6); both imported from West Friesland by GERRIT S. MILLER, 1869.

No. 15. Juno.

White and black; calved March 4, 1872; bred by and the property of GERRIT S. MILLER, Peterboro', Madison County, N. Y. Sire, Hollander (20); imported from West Friesland by GERRIT S. MILLER, 1869; dam, Crown Princess (6); imported from West Friesland by GERRIT S. MILLER, 1869.

No. 16. Lady Andover.

Black and white; calved December 31, 1869; bred by and the property of W. A. RUSSELL, Lawrence, Mass. Sire, Zuider Zee 2d (57); imported from North Holland by WINTHROP W. CHENERY, 1861; dam, Midwould 4th (26); by Hollander (19); grandam, Lady Midwould (17); imported from North Holland by WINTHROP W. CHENERY, 1861.

No. 17. Lady Midwould.

Black and white; calved in 1859; bred in North Holland; thence imported by and the property of WINTHROP W. CHENERY, Belmont, Mass., 1861.

No. 18. Lady Oldenburg.

Black and white; calved in 18—; bred in the Duchy of Oldenburg; imported by Dr. HATCH; the property of W. A. RUSSELL, Lawrence, Mass.

No. 19. Lady Rutten.

Black and white; calved in 1865; bred in North Holland; thence imported by WINTHROP W. CHENERY, Belmont, Mass., 1857. Slaughtered in 1860, leaving no pure-bred progeny, excepting her son, 2d Dutchman (37).

No. 20. Lady Van Tromp.

White and black; calved January 8, 1872; bred by and the property of JOHN CUMMINGS, Woburn, Mass. Sire, 2d Duke of Belmont (36); grandsire, Van Tromp (50); imported from North Holland by WINTHROP W. CHENERY, 1861. Dam, Opperdoes 14th (42), by Van Tromp (50); grandam, Maid of Opperdoes (22); imported from North Hollandby WINTHROP W. CHENERY, 1861.

No. 21. Maid of Holstein.

Black and white; calved March 12, 1871; bred by and the property of THOMAS B. WALES, Jr., South Framingham, Mass. Sire, Opperdoes 7th (32); grandsire, Van Tromp (50); imported from North Holland by WINTHROP W. CHENERY, 1861. Dam, Maid of Opperdoes (22); also imported from North Holland by WINTHROP W. CHENERY, 1861.

No. 22. Maid of Opperdoes.

Black and white; calved in 1859; bred in North Holland; thence imported by WINTHROP W. CHENERY, Belmont, Mass., 1861; the property of THOMAS B. WALES, Jr., South Framingham, Mass.

No. 23. Maud.

Black and white; calved September 2d, 1871; bred by THOMAS B. WALES, Jr., South Framingham, Mass.; the property of JOHN H. COMER, Goshen, N. Y. Sire, Opperdoes 7th (32); grandsire, Van Tromp (50); imported from North Holland by WINTHROP W. CHENERY, 1861. Dam, Opperdoes 10th (41), by Opperdoes 4th (29); grandam, Opperdoes 2d (38), by 2d Dutchman (37); g. grandam, Maid of Opperdoes (22); imported from North Holland by WINTHROP W. CHENERY, 1861.

No. 24. Meenie.

Black and white; calved April 14, 1871; bred by and the property of CHARLES HOUGHTON, Putney, Vt. Sire, Amsterdam (1); grandsire, Van Tromp (50); imported from North Holland by WINTHROP W. CHENERY, 1861. Dam, Texelaar 8th (55); by Zuider Zee 2d (57); grandam, Texelaar (51); also imported from North Holland by WINTHROP W. CHENERY, 1861.

No. 25. Midwould 2d.

Black, with white spots; calved October 15, 1862; bred by and the property of WINTHROP W. CHENERY, Belmont, Mass. Sire, 2d Dutchman (37); grandsire, Dutchman (7); imported from North Holland by WINTHROP W. CHENERY, 1857. Dam, Lady Midwould (17); imported from North Holland by WINTHROP W. CHENERY, 1861.

No. 26. Midwould 4th.

Black and white; calved July 1, 1864; bred by WINTHROP W. CHENERY, Belmont, Mass.; the property of W. A. RUSSELL, Lawrence, Mass. Sire, Hollander (19); dam, Lady Midwould (17); both imported from North Holland by WINTHROP W. CHENERY, 1861.

No. 27. Midwould 5th.

Black and white; calved June 15, 1865; bred by WINTHROP W. CHENERY, Belmont, Mass.; the property of CHARLES HOUGHTON, Putney, Vt. Sire, Zuider Zee 2d (57); imported from North Holland by WINTHROP W. CHENERY, 1861. Dam, Midwould 2d (25), by 2d Dutchman (37); grandam, Lady Midwould (17); imported from North Holland by WINTHROP W. CHENERY, 1861.

No. 28. Midwould 5th.

Black and white; calved June 13, 1867; bred by and the property of W. A. RUSSELL, Lawrence, Mass. Sire, Zuider Zee 2d (57); imported from North Holland by WINTHROP W. CHENERY, 1861. Dam, Midwould 4th (26), by Hollander (19); grandam, Lady Midwould (17); imported from North Holland by WINTHROP W. CHENERY, 1861.

No. 29. Midwould 6th.

White and black; calved November 22, 1865; bred by WINTHROP W. CHENERY, Belmont, Mass.; the property of CHARLES HOUGHTON, Putney, Vt. Sire, Zuider Zee 2d (57); imported from North Holland by WINTHROP W. CHENERY, 1861. Dam, Lady Midwould (17); also imported from North Holland by WINTHROP W. CHENERY, 1861.

No. 30. Midwould 6th.

Black and white; calved April 6, 1868; bred by and the property of W. A. RUSSELL, Lawrence, Mass. Sire, Zuider Zee 2d (57); imported from North Holland by WINTHROP W. CHENERY, 1861. Dam, Midwould 4th (26), by Hollander (19); grandam, Lady Midwould (17); imported from North Holland by WINTHROP W. CHENERY, 1861.

No. 31. Midwould 8th.

Black, with a little white; calved February 1, 1868; bred by WINTHROP W. CHENERY, Belmont, Mass.; the property of C. C. WALWORTH, Monticello, Iowa. Sire, Zuider Zee 4th (59); grandsire, Van Tromp (50); imported from North Holland by WINTHROP W. CHENERY, 1861. Dam, Midwould 5th (27); by Zuider Zee 2d (57); grandam, Midwould 2d (25), by 2d Dutchman (37); g. grandam, Lady Midwould (17); imported from North Holland by WINTHROP W. CHENERY, 1861.

No. 32. Midwould 9th.

Black and white; calved May 23, 1868; bred by and the property of WINTHROP W. CHENERY, Belmont, Mass. Sire, Van Tromp (50); Dam, Lady Midwould (17); both imported from North Holland by WINTHROP W. CHENERY, 1861.

No. 33. Midwould 10th.

White and black; calved June 17, 1868; bred by WINTHROP W. CHENERY, Belmont, Mass.; the property of RUFUS WATERMAN, Jr., Norwich, Conn. Sire, Opperdoes 7th (32); grandsire, Van Tromp (50); imported from North Holland by WINTHROP W. CHENERY, 1861. Dam, Midwould 6th (29), by Zuider Zee 2d (57); grandam, Lady Midwould (17); imported from North Holland by WINTHROP W. CHENERY, 1861.

No. 34. Midwould 13th.

Black, with white marks; calved March 17, 1871; bred by WINTHROP W. CHENERY, Belmont, Mass.; the property of B. E. STEWART, North Yamhall, Yamhill Co., Oregon. Sire, Texelaar 6th (44); grandsire, Zuider Zee 2d (57); imported from North Holland by WINTHROP W. CHENERY, 1861. Dam, Lady Midwould, (17); imported from North Holland by WINTHROP W. CHENERY, 1861.

No. 35. Midwould 15th.

Black and white; calved May 17, 1871; bred by WINTHROP W. CHENERY, Belmont, Mass.; the property of the ONEIDA COMMUNITY, Oneida, N. Y. Sire, Texelaar 6th (44); grandsire, Zuider Zee 2d (57); imported from North Holland by WINTHROP W. CHENERY, 1861. Dam, Midwould 2d (25), by 2d Dutchman (37); grandam, Lady Midwould (17); imported from North Holland by WINTHROP W. CHENERY, 1861.

No. 36. Midwould 16th.

White and black; calved August 4, 1871; bred by WINTHROP W. CHENERY, Belmont, Mass.; the property of JOHN H. COMER, Goshen, N. Y. Sire, Texelaar 6th (44); grandsire, Zuider Zee 2d (57); imported from North Holland by WINTHROP W. CHENERY, 1861. Dam, Midwould 10th (33), by Opperdoes 7th (32); grandam, Midwould 6th (29), by Zuider Zee 2d (57); g. grandam, Lady Midwould (17); imported from North Holland by WINTHROP W. CHENERY, 1861.

No. 37. Midwould 17th.

Black and white; calved March 23, 1872; bred by and the property of WINTHROP W. CHENERY, Belmont, Mass. Sire, Duke of Belmont (5);

grandsire, Van Tromp (50); imported from North Holland by WINTHROP W. CHENERY, 1861. Dam, Lady Midwould (17); imported from North Holland by WINTHROP W. CHENERY, 1861.

No. 38. Opperdoes 2d.

White and black; calved Feb. 15, 1863; bred by WINTHROP W. CHENERY, Belmont, Mass.; the property of JOHN H. COMER, Goshen, N. Y. Sire, 2d Dutchman (37); grandsire, Dutchman (7); imported from North Holland by WINTHROP W. CHENERY, 1857. Dam, Maid of Opperdoes (22); imported from North Holland by WINTHROP W. CHENERY, 1861.

No. 39. Opperdoes 3d.

Black and white; calved Feb. 25, 1864; bred by and the property of WINTHROP W. CHENERY, Belmont, Mass. Sire, Hollander (19); dam, Maid of Opperdoes (22); both imported from North Holland by WINTHROP W. CHENERY, 1861.

No. 40. Opperdoes 8th.

Black and white; calved March 9, 1867; bred by WINTHROP W. CHENERY, Belmont, Mass.; the property of S. B. EMERSON, Mountain View, Santa Clara Co., California. Sire, Van Tromp (50); dam, Maid of Opperdoes (22); both imported from North Holland by WINTHROP W. CHENERY, 1861.

No. 41. Opperdoes 10th.

White and black; calved Oct. 17, 1867; bred by WINTHROP W. CHENERY, Belmont, Mass.; the property of JOHN H. COMER, Goshen, N. Y. Sire, Opperdoes 4th (29); grandsire, Zuider Zee 2d (57); imported from North Holland by WINTHROP W. CHENERY, 1861. Dam, Opperdoes 2d (38), by 2d Dutchman (37); grandam, Maid of Opperdoes (22); imported from North Holland by WINTHROP W. CHENERY, 1861.

No. 42. Opperdoes 14th.

White and black; calved March 14, 1869; bred by WINTHROP W. CHENERY, Belmont, Mass.; the property of JOHN CUMMINGS, Woburn, Mass. Sire, Van Tromp (50); dam, Maid of Opperdoes (22); both imported from North Holland by WINTHROP W. CHENERY, 1861.

No. 43. Opperdoes 15th.

White, with black spots; calved Nov. 12, 1869; bred by WINTHROP W. CHENERY, Belmont, Mass.; the property of WILLIAM STECKEL, Doylestown, Bucks Co., Penn. Sire, Texelaar 6th (44); grandsire, Zuider Zee 2d (57); imported from North Holland by WINTHROP W. CHENERY, 1861. Dam, Opperdoes 2d (38), by 2d Dutchman (37); grandam, Maid of Opperdoes (22); imported from North Holland by WINTHROP W. CHENERY, 1861.

No. 44. Opperdoes 16th.

Black and white; calved March 4, 1870; bred by WINTHROP W. CHENERY, Belmont, Mass.; the property of ONEIDA COMMUNITY, Oneida, N. Y. Sire, Van Tromp (50); dam, Maid of Opperdoes (22); both imported from North Holland by WINTHROP W. CHENERY, 1861.

No. 45. Opperdoes 18th.

Black and white; calved April 2, 1871; bred by WINTHROP W. CHENERY, Belmont, Mass.; the property of JOHN CUMMINGS, Woburn, Mass. Sire, Texelaar 6th (44); grandsire, Zuider Zee 2d (57); imported from North Holland by WINTHROP W. CHENERY, 1861. Dam, Opperdoes 2d (38), by 2d Dutchman (37); grandam, Maid of Opperdoes (22); imported from North Holland by WINTHROP W. CHENERY, 1861.

No. 46. Opperdoes 19th.

Black and white; calved May 5, 1871; bred by and the property of WINTHROP W. CHENERY, Belmont, Mass. Sire, Texelaar 6th (44); grandsire, Zuider Zee 2d (57); imported from North Holland by WINTHROP W. CHENERY, 1861. Dam, Opperdoes 3d (39), by Hollander (19); grandam, Maid of Opperdoes (22); imported from North Holland by WINTHROP W. CHENERY, 1861.

No. 47. Prairie Maid.

White and black; calved October 19, 1871; bred by and the property of C. C. WALWORTH, Monticello, Iowa. Sire, Texelaar 6th (44); grandsire, Zuider Zee 2d (57); imported from North Holland by WINTHROP W. CHENERY, 1861. Dam, Zuider Zee 3d (63), by Hollander (19); grandam, Zuider Zee (62); imported from North Holland by WINTHROP W. CHENERY, 1861.

No. 48. Queen of the West.

White and black; calved June, 1871; bred by and the property of C. C. WALWORTH, Monticello, Iowa. Sire, Midwould 4th (26); grandsire, Duke of Holstein (6); g. grandsire, Hollander (19); imported from North Holland by WINTHROP W. CHENERY, 1861. Dam, Midwould 8th (31), by Zuider Zee 4th (59); grandam, Midwould 5th (27), by Zuider Zee 2d (57); g. grandam, Midwould 2d (25), by 2d Dutchman (37); g. g. grandam, Lady Midwould (17); imported from North Holland by WINTHROP W. CHENERY, 1861.

No. 49. Santa Clara.

Black and white; calved November 20, 1870; bred by and the property of S. B. EMERSON, Mountain View, Santa Clara County, Cal. Sire, Opperdoes 4th (29); grandsire, Zuider Zee 2d (57); imported from North Holland by WINTHROP W. CHENERY, Belmont, Mass., 1861. Dam, Opperdoes 8th (40), by Van Tromp (50); grandam, Maid of Opperdoes (22); imported from North Holland by WINTHROP W. CHENERY, 1861.

No. 50. Snowflake.

White and black; calved October 27, 1871; bred by and the property of GERRIT S. MILLER, Peterboro', N. Y. Sire, Hamilcar (17); dam, Agoo (1); both imported from West Friesland by GERRIT S. MILLER, 1869.

No. 51. Texelaar.

Black and white; calved in 1859; bred in North Holland; thence imported by and the property of WINTHROP W. CHENERY, Belmont, Mass., 1861.

No. 52. Texelaar 3d.

Black and white; calved April 2, 1863; bred by WINTHROP W. CHENERY, Belmont, Mass.; the property of HENRY WATERMAN, North Kingston, R. I. Sire, 2d Dutchman (37); grandsire, Dutchman (7); imported from North Holland by WINTHROP W. CHENERY, 1857. Dam, Texelaar (51); imported from North Holland by WINTHROP W. CHENERY, 1861.

No. 53. Texelaar 5th.

White, with black spots; calved May 15, 1865; bred by and the property of WINTHROP W. CHENERY, Belmont, Mass. Sire, Hollander (19); dam, Texelaar (51); both imported from North Holland by WINTHROP W. CHENERY, 1861.

No. 54. Texelaar 7th.

White, with black spots; calved May 25, 1866; bred by and the property of WINTHROP W. CHENERY, Belmont, Mass. Sire, Zuider Zee 2d (57); dam, Texelaar (51); both imported from North Holland by WINTHROP W. CHENERY, 1861.

No. 55. Texelaar 8th.

Black and white; calved May 25, 1866; bred by WINTHROP W. CHENERY, Belmont, Mass.; the property of CHARLES HOUGHTON, Putney, Vt. Sire, Zuider Zee 2d (57); dam, Texelaar (51); both imported from North Holland by WINTHROP W. CHENERY, 1861.

No. 56. Texelaar 9th.

Black and white; calved April 28, 1867; bred by WINTHROP W. CHENERY, Belmont, Mass.; the property of WILLIAM STECKEL, Doylestown, Bucks County, Penn. Sire, Van Tromp (50); imported from North Holland by WINTHROP W. CHENERY, 1861. Dam, Texelaar 3d (52), by 2d Dutchman (37); grandam, Texelaar (51); imported from North Holland by WINTHROP W. CHENERY, 1861.

No. 57. Texelaar 10th.

Black and white; calved July 3, 1867; bred by and the property of WINTHROP W. CHENERY, Belmont, Mass. Sire, Opperdoes 4th (29); grandsire, Zuider Zee 2d (57); imported from North Holland by WINTHROP W. CHENERY, 1861. Dam, Texelaar (51); imported from North Holland by WINTHROP W. CHENERY, 1861.

No. 58. Texelaar 11th.

Black and white; calved March 11, 1868; bred by and the property of WINTHROP W. CHENERY, Belmont, Mass. Sire, Van Tromp (50); imported from North Holland by WINTHROP W. CHENERY, 1861. Dam, Texelaar 3d (52), by 2d Dutchman (37); grandam, Texelaar (51); also imported from North Holland by WINTHROP W. CHENERY, 1861.

No. 59. Texelaar 12th.

Black and white; calved June 20, 1869; bred by WINTHROP W. CHENERY, Belmont, Mass.; the property of JOHN H. COMER, Goshen, N. Y. Sire, 3d Dutchman (46); g. sire, 2d Dutchman (37); g. g. sire, Dutchman (7); imported from North Holland by WINTHROP W. CHENERY, 1857. Dam, Texelaar 8th (55), by Zuider Zee 2d (57); grandam, Texelaar (51); imported from North Holland by WINTHROP W. CHENERY, 1861.

No. 60. Texelaar 14th.

Black, with white marks; calved April 1, 1870; bred by WINTHROP W. CHENERY, Belmont, Mass.; the property of HENRY WATERMAN, North Kingston, R. I. Sire, Opperdoes 4th (29); grandsire, Zuider Zee 2d (57); imported from North Holland by WINTHROP W. CHENERY, 1861. Dam, Texelaar 7th (54), by Zuider Zee 2d (57); grandam, Texelaar (51); imported from North Holland by WINTHROP W. CHENERY, 1861.

No. 61. Topsey.

Black, with very little white; calved June 14, 1871; bred by and the property of GERRIT S. MILLER, Peterboro', N. Y. Sire, Hollander (20); dam, Dowager (7); both imported from West Friesland by GERRIT S. MILLER, 1869.

No. 62. Zuider Zee.

Black and white; calved in 1859; bred in North Holland; thence imported by and the property of WINTHROP W. CHENERY, Belmont, Mass., 1861.

No. 63. Zuider Zee 3d.

White, with black spots; calved March 14, 1864; bred by WINTHROP W. CHENERY, Belmont, Mass.; the property of C. C. WALWORTH, Monticello, Iowa. Sire, Hollander (19); dam, Zuider Zee (62); both imported from North Holland by WINTHROP W. CHENERY, 1861.

No. 64. Zuider Zee 5th.

Black and white; calved May 5, 1867; bred by and the property of WINTHROP W. CHENERY, Belmont, Mass. Sire, Van Tromp (50); dam, Zuider Zee (62); both imported from North Holland by WINTHROP W. CHENERY, 1861.

No. 65. Zuider Zee 9th.

Black, with white marks; calved March 4, 1870; bred by WINTHROP W. CHENERY, Belmont, Mass.; the property of THOMAS B. WALES, Jr., South Framingham, Mass. Sire, Van Tromp (50); dam, Zuider Zee (62); both imported from North Holland by WINTHROP W. CHENERY, 1861.

No. 66. Zuider Zee 10th.

White and black; calved July 24, 1870; bred by WINTHROP W. CHENERY, Belmont, Mass.; the property of RUFUS WATERMAN, Jr., Norwich, Conn. Sire, 3d Dutchman (46); grandsire, 2d Dutchman (37); g. g. sire, Dutchman (7); imported from North Holland by WINTHROP W. CHENERY, 1857. Dam, Zuider Zee 5th (64), by Van Tromp (50); grandam, Zuider Zee (62); imported from North Holland by WINTHROP W. CHENERY, 1861.

No. 67. Zuider Zee 12th.

Black and white; calved October 4, 1871; bred by and the property of WINTHROP W. CHENERY, Belmont, Mass. Sire, Duke of Holstein (6); grandsire, Hollander (19); imported from North Holland by WINTHROP W. CHENERY, 1861. Dam, Zuider Zee 5th (64), by Van Tromp (50); grandam, Zuider Zee (62); imported from North Holland by WINTHROP W. CHENERY, 1861.

INDEX.

Holstein Heifer, OPPERDOES 3d, at 10 Months Old. Bred by Winthrop W. Chenery.

Holstein Cow, **LADY MIDWOULD.** Imported from North Holland by Winthrop W. Chenery.

Posterior View of **LADY MIDWOULD**, showing Milk Escutcheon.

Holstein Cow, **TEXELAAR.** Imported from North Holland by Winthrop W. Chenery.

Holstein Bull, **VAN TROMP.** Imported from North Holland by Winthrop W. Chenery.

Holstein Bull, HOLLANDER. Imported from North Holland by Winthrop W. Chenery.

Made in the USA
Columbia, SC
26 January 2025